Maths Skills *for GCSE*

Computer Science

Alison Page

OXFORD
UNIVERSITY PRESS

Great Clarendon Street, Oxford, OX2 6DP, United Kingdom

Oxford University Press is a department of the University of Oxford. It furthers the University's objective of excellence in research, scholarship, and education by publishing worldwide. Oxford is a registered trade mark of Oxford University Press in the UK and in certain other countries

British Library Cataloguing in Publication Data
Data available

978-0-19-843791-8

10 9 8 7 6 5 4 3 2 1

Printed in India by Manipal Technologies Limited

Acknowledgements

Cover: Rauf Aliyev/Shutterstock

Artwork by Thomson Digital

Contents

How to use this book

This workbook has been written to support the development of key mathematics skills required to achieve success in your GCSE Computer Science course. It has been devised and written by teachers and the practice questions included reflect the **exam-tested content** for AQA, OCR, Edexcel and Eduqas specifications.

The workbook is structured into chapters, with each chapter relating to an area of computer science. Then, each topic covers a mathematical skill or skills that you may need to practise. Each topic offers the following features:

Maths Skills for GCSE Computer Science

4.2 Turn a bitstream into an image

Make an image

A monochrome image can be created from the 1s and 0s of a bitstream. ❶

- Put the 1s and 0s into a pixel grid.
- Replace 1s with white and 0s with black.

The only other information needed is the width and height of the pixel grid.

WORKED EXAMPLE

Interpret the following bit stream as a bitmap image in a 4 × 4 grid. ❷

1101110100001101

A 4 × 4 grid looks like this:

There are 16 pixels in the grid, and 16 bits in the bitstream. Starting in the top left hand corner, copy the 1s and 0s of the bitstream into the grid. It may be helpful to split them into groups of four to match the rows of the grid.

1101
1101
0000
1101

Each group of four can be copied into one row of the grid.

1	1	0	1
1	1	0	1
0	0	0	0
1	1	0	1

The 1s become white pixels and the 0s become black pixels.

1	1	0	1
1	1	0	1
0	0	0	0
1	1	0	1

This is the completed image.

PRACTICE QUESTION ❹

1 Convert the following bitstreams into bitmap images.

a 1100110000110011 in a 4 × 4 grid

b 1010100000101010000010101 in a 5 × 5 grid

c 110011 101101 111101 110011 110111 111111 110111 in a 6 × 7 grid

44

4 Digital images

WORKED EXAMPLE

Convert the following bitstream to a bitmap image in a 6 × 8 grid.

111111110011111011111011100011101011100001111111

The grid is described as 6 × 8. This means the width is 6 and the height is 8.

Add the pixels to the grid starting in the top left hand corner. Then shade in the pixels marked with a 0.

1	1	1	1	1	1
1	1	0	0	1	1
1	1	1	0	1	1
1	1	1	0	1	1
1	0	0	0	1	1
1	0	1	0	1	1
1	0	0	0	0	1
1	1	1	1	1	1

1	1	1	1	1	1
1	1	0	0	1	1
1	1	1	0	1	1
1	1	1	0	1	1
1	0	0	0	1	1
1	0	1	0	1	1
1	0	0	0	0	1
1	1	1	1	1	1

The design shows the lower case letter d.

PRACTICE QUESTION

2 A font designer creates some simple letter designs as bitstreams. In each case the pixel grid was 8 × 8.

Give the letters shown in the bitmaps represented by the following bitstreams.

a 11111111 / 11000111 / 10010011 / 10111011 / 10000011 / 10111111 / 10010011 / 11000111

b 11111111 / 10000011 / 10111111 / 10111111 / 10000111 / 10111111 / 10111111 / 10000011

c 11111111 / 11000111 / 10010011 / 10111011 / 10111011 / 10111011 / 10010011 / 11000111

d 11111111 / 10000111 / 11011011 / 11011011 / 11000111 / 11011011 / 11011011 / 10000111

Extra bits

Sometimes computer systems use additional bits to record the bitmap. For example, there might be extra bits 'padding out' the start and end of each line. In cases where this is important, the information you need will be provided.

STRETCH YOURSELF! ❺

Design some letters in a font of your own design and represent them as a bitstream.

REMEMBER: To change a bitstream into an image:

1 find the size of the grid and draw it
2 insert the series of 1s and 0s into the grid
3 shade cells containing 0. ❸

45

❶ ***Opening paragraph*** outlines the mathematical skill or skills covered within the topic.

❷ ***Worked example*** – Annotated worked examples offer a step-by-step breakdown of how to answer questions related to each topic.

❸ ***Remember*** is a useful box that will offer you tips, hints and other snippets of useful information.

❹ ***Summary questions*** are ramped in terms of difficulty and all answers are available at www.oxfordsecondary.co.uk

❺ ***Stretch yourself*** – some of the topics may also contain a few more difficult questions to stretch your mathematical knowledge and understanding.

1.1 Decimal (base 10)

Decimal digits

The number system that we use in everyday life is called decimal. There are ten digits that can be used to make a decimal number. Each digit stands for a different number value. The ten digits are:

0 1 2 3 4 5 6 7 8 9

Because decimal uses ten digits, we call it 'base 10'. Another name for this number system is denary.

Counting in base 10

Although there are only ten digits, the decimal system can represent far more than ten numbers. In fact, any whole number value can be represented using the decimal system. This is because digits have different values that depend on their position in the number.

The first ten decimal values are represented by the digits 0 to 9. Counting up to nine reaches the largest decimal digit. To make the next number:

- reset the 9 back to 0
- put 1 in the tens column.

The number 99 has the largest available decimal digit for both tens and ones. To make the next number:

- reset the tens and ones back to 0
- put 1 in the hundreds column.

Nine is the largest digit in decimal numbers. When counting beyond nine, reset to zero and increase the next position by one.

Place values

The value of each digit in a decimal number is based on its position in that number. Think of the four-digit number 1529. The value of each digit depends on its position in the number.

Thousands	Hundreds	Tens	Ones
1	5	2	9

The 1 is in the thousands column so it stands for 1000. The 5 is in the hundreds column so it stands for 500. The 2 is in the tens column so it stands for 20. The 9 is in the ones column so it stands for 9. The ones column is sometimes called the units column.

The key features of the decimal position grid are as follows.

- The position values start at one, on the right of the grid.
- The values increase from right to left.
- Each position value is ten times bigger than the previous position value.

PRACTICE QUESTION

1 Give the value of the digit 6 in the decimal number 106 709.

a six million **b** six thousand

c six hundred **d** six hundred thousand

WORKED EXAMPLE

Give the total value of the decimal number 106 709.

To calculate the value of a number in any number system:

- multiply each digit value by its position value to give the value represented by that digit
- add together the value represented by each digit to give the total value of the number.

1 Put the digits into the grid to find their position values.

Hundred thousands	Ten thousands	Thousands	Hundreds	Tens	Ones
1	0	6	7	0	9

2 Multiply the value of each digit by its position value. In this case the digit 6 in the thousands column stands for 6000. Add the values of all the digits together to give the total value of the number.

Why use different number bases?

In this book you will learn about several number bases and why they are important in computer science:

- *decimal* (base 10) – because the earliest counting systems used fingers, and people have ten fingers
- *binary* (base 2) – because signals within the computer are stored in binary form
- *hexadecimal* (base 16) – because converting between binary and hexadecimal is easy.

You will learn more about binary and hexadecimal in the rest of this chapter.

PRACTICE QUESTION

2 a In the decimal number 192 837, state which digit represents the largest value.

b In the decimal number 456 789, give the value that is represented by the digit 4.

c Give the largest whole number that can be made using each decimal digit once only. Remember to use the digit 0 as well as the others.

d Give the smallest whole number that can be made using each decimal digit once only.

1.2 Binary (base 2)

Binary digits (bits)

Computer scientists often work with binary numbers. 'Binary' means base 2. There are two digits that can be used to make a binary number:

0 1

Binary digits are also called bits. Bit is short for binary digit.

Any number that can be represented in decimal can also be represented in binary.

Counting in base 2

Although there are only two values for each bit, 1 and 0, the binary system can use these two values to represent any whole number.

The first two binary values are represented by single bits:

0
1

There are now no more binary digits available to you to increase the value of this column. To move to the next number value, you should reset the units back to the 0 and put 1 in the next column to the left. The sequence is now:

0
1
10

To add one, increase the digit on the right from 0 to 1:

0
1
10
11

Again, you then have no more digits to increase the value of this column. Once more, reset the digits to 0 and put 1 in the next column to the left:

0
1
10
11
100

This process of resetting to 0 and adding a 1 in the column to the left repeats. The sequence of binary numbers from 0 upwards is shown in Table 1.

Table 1 Binary numbers representing values from zero to eight

Binary number	Decimal number
0	0
1	1
10	2
11	3
100	4
101	5
110	6
111	7
1000	8

PRACTICE QUESTION

1 Extend the sequence in Table 1 to show all binary numbers up to a value of 15.

Place values in binary

Each bit in the binary number has a different value based on its position.

The position on the right has the value 1. The values increase from right to left. Each position value is double the previous position value. This grid shows the first eight position values.

128	64	32	16	8	4	2	1

To find the value of a binary number, place the bits of the number into the binary grid, following these rules.

- Each position must only hold one bit – a 1 or a 0.
- The bits should be filled in starting on the right of the grid. Any empty spaces on the left of the grid should be filled with a 0.

For example, placing the number 1001 in the grid would look like this:

128	64	32	16	8	4	2	1
0	0	0	0	1	0	0	1

Then add together the value of every position where there is a 1.

8 + 1 = 9

NOTE: Binary numbers are often written using 8 bits. To make the number easier to read, the bits are split into two groups of four, for example, 0010 1101.

WORKED EXAMPLE

Calculate the decimal value of the binary number 0001 0100.

To work this out, enter the 1s and 0s into the grid that shows position values.

128	64	32	16	8	4	2	1
0	0	0	1	0	1	0	0

Find the value of every 1 in the number.

There is a 1 in the 16 position.

There is a 1 in the 4 position.

Add these position values together.

16 + 4 = 20

The value of the binary number 0001 0100 in decimal form is 20.

PRACTICE QUESTIONS

2 Calculate the decimal value of the following binary numbers.

a 0010 0010 **b** 0001 0100 **c** 0100 0001 **d** 1000 1000

3 Calculate the decimal value of each of these binary numbers. The numbers include more 1s so you will need to add more values together.

a 0110 1110 **b** 1101 0110 **c** 0110 1111 **d** 1011 1101

REMEMBER: To convert binary to decimal, place the bits in the binary grid. Add the values of all the positions where there is a 1.

NOTE: You do not pronounce binary numbers as if they were decimal numbers. For example, the binary number 10 is not pronounced 'ten'. To name a binary number, name its digits in order. You might say 'one oh' or 'one zero'.

1.3 Converting between decimal and binary

Convert binary to decimal

To convert a binary number to decimal – add up the binary position values as follows.

- Place the bits of the binary number into the binary position grid.

128	64	32	16	8	4	2	1

There is no need to memorise the position values. Simply start at 1 on the right and double the value each time as you move to the left.

- Remember to place the bits as far to the right as possible, and fill any empty spaces on the right hand side with zeros.
- Finally, add together the values of all positions where there is a 1.

PRACTICE QUESTION

1 Convert these binary numbers to decimal.

a 0011 0011

b 0101 0110

c 0110 0111

d 1001 1001

Convert decimal to binary

To convert a decimal number to binary – subtract the binary position values by:

- subtracting the largest value possible without making a negative result
- putting a 1 under each value that you subtract in the grid
- continuing to subtract position values until the result is 0.

WORKED EXAMPLE

Convert the decimal number 50 to binary.

Look at the grid of binary position values. The largest number that can be subtracted from 50 is 32. Place a 1 under that value.

128	64	32	16	8	4	2	1
		1					

50 – 32 = 18

The result is 18. The largest number that can be subtracted from 18 is 16. Place a 1 under that value.

128	64	32	16	8	4	2	1
		1	1				

18 – 16 = 2

The result is 2. Place a 1 under that value.

128	64	32	16	8	4	2	1
		1	1			1	

Subtracting 2 reduces the result to 0 so the process is over.
Put zeros into all the remaining empty positions.

128	64	32	16	8	4	2	1
0	0	1	1	0	0	1	0

To get the final answer, show the bits without the grid. To make them easier to read you can split them into two groups of four:

0011 0010

PRACTICE QUESTIONS

2 Convert these decimal numbers to binary.

a 23

b 99

c 60

d 88

3 Convert these larger decimal numbers to binary.

a 154

b 201

c 222

d 198

REMEMBER: To convert decimal to binary, subtract the values in the binary grid. Start with the highest values. Only subtract if the result is greater than 0. Put a 1 into every position where you have subtracted a value.

1.4 Hexadecimal (base 16)

Hexadecimal digits

Hexadecimal is base 16. Hexadecimal numbers are made from 16 digits.

1 2 3 4 5 6 7 8 9 A B C D E F

The first ten digits (0 to 9) have the same value as decimal digits. The next six digits (A to F) represent the numbers 10 to 15.

Counting in base 16

The 16 hexadecimal digits count up from 0 to 15, as shown in Table 1.

Table 1 Hexadecimal digits up to a value of 15

Decimal value	Hexadecimal digit
0	0
1	1
2	2
3	3
4	4
5	5
6	6
7	7
8	8
9	9
10	A
11	B
12	C
13	D
14	E
15	F

Counting beyond 15

F is the largest hexadecimal digit. It represents the decimal value 15.
To count beyond F:

- reset the ones column to 0
- add 1 to the next column.

So the decimal number 16 is shown as 10 in hexadecimal, 17 in decimal is shown as 11 in hexadecimal, and so on.

PRACTICE QUESTION

1 Extend Table 1 to give the hexadecimal numbers that represent the decimal values from 16 to 31.

Position values

The decimal values from 0 to 255 can be represented by two-digit hexadecimal numbers. The value of each hexadecimal digit is found by multiplying the digit value by the position value. The total value of the number is found by adding the digit values together.

With a two-digit hexadecimal number this means:

- multiplying the digit on the left by 16
- adding the value of the digit on the right.

WORKED EXAMPLE

Calculate the decimal value of the hexadecimal number 34.

Place the digits of the number in the hexadecimal grid.

Sixteens	Ones
3	4

The digit on the left is 3. Multiply that by 16 to give the value 48.

$(3 \times 16) = 48$

Then add the digit on the right.

$48 + 4 = 52$

So 34 in hexadecimal is 52 in decimal.

PRACTICE QUESTION

2 Convert these hexadecimal numbers into decimal numbers.

a 21 **b** 45 **c** 60 **d** 77

NOTE: Some hexadecimal digits are letters of the alphabet but not all hexadecimal numbers include digits of this type. You can't tell whether a number is decimal or hexadecimal just by checking whether it has letters in it. That is why it is always important to state what number base you are using.

WORKED EXAMPLE

Calculate the decimal value of the hexadecimal number A4.

Some hexadecimal numbers include the digits that stand for the values 10 to 15. The same method is used but you must work out the value of the digit before you multiply it by the position value.

Sixteens	Ones
A	4

The digit on the left is A. This is equivalent to the decimal value 10. Multiply that by 16 to give the value 160.

$10 \times 16 = 160$

Then add the digit on the right.

$160 + 4 = 164$

The hexadecimal number A4 is 164 in decimal.

PRACTICE QUESTION

3 Give the decimal equivalent of these hexadecimal numbers.

a 2B **b** C5 **c** E0 **d** 7F

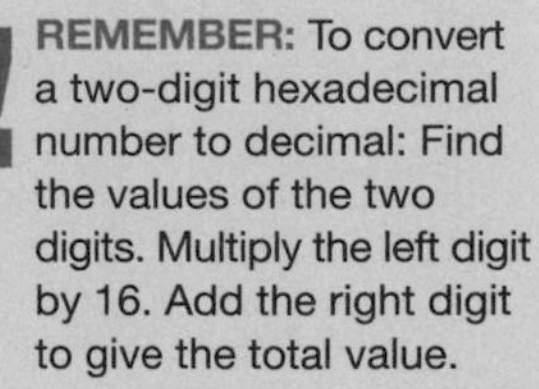

REMEMBER: To convert a two-digit hexadecimal number to decimal: Find the values of the two digits. Multiply the left digit by 16. Add the right digit to give the total value.

1.5 Converting between decimal and hexadecimal

Converting hexadecimal to decimal

Hexadecimal means base 16. In a two-digit hexadecimal number the digit on the right stands for the number of ones. The number on the left stands for the number of sixteens. You can remember this by putting the hexadecimal number into the position grid.

Sixteens	Ones
2	B

You then multiply the value of each digit by its position. Remember B has the value 11.

$2 \times 16 = 32$

$B \times 1 = 11 \times 1 = 11$

Add the two numbers.

$32 + 11 = 43$

So 2B in hexadecimal is 43 in decimal.

Converting decimal to hexadecimal

Decimal numbers between 0 and 15 can be represented by a single hexadecimal digit. See Table 1 in Topic 1.4 to remind yourself of the 16 hexadecimal digits and their values.

To turn a decimal number larger than 15 into hexadecimal, work out two values:

- how many sixteens there are in the number – this amount goes in the sixteens position
- the remainder – this value goes in the ones position.

The 16 times table (Table 1) can help with this task.

Table 1 Sixteen times table for converting decimal to hexadecimal numbers

Hexadecimal digit in the sixteens position	Multiply digit value by 16	Decimal value of the digit in the sixteens position
0	16 × 0	0
1	16 × 1	16
2	16 × 2	32
3	16 × 3	48
4	16 × 4	64
5	16 × 5	80
6	16 × 6	96
7	16 × 7	112
8	16 × 8	128
9	16 × 9	144
A	16 × 10	160
B	16 × 11	176
C	16 × 12	192
D	16 × 13	208
E	16 × 14	224
F	16 × 15	240

WORKED EXAMPLE

Give the hexadecimal equivalent of the decimal number 88.

The first step is to work out how many sixteens there are in 88. You can work out the division yourself, use a calculator, or look at Table 1. You must find the nearest number smaller than the target value.

$5 \times 16 = 80$

The digit 5 in the sixteens position has the value 80. Put a 5 in the sixteens position. Then subtract its value from the original number.

$88 - 80 = 8$

This gives the remainder, which is 8. The remainder goes in the ones position. Here is the hexadecimal number drawn in the position grid.

Sixteens	Ones
5	8

You can then read the number from the grid. The decimal number 88 is equivalent to the hexadecimal number 58.

PRACTICE QUESTION

1 Convert the following decimal numbers to hexadecimal.

a 39 **b** 115 **c** 80 **d** 40

WORKED EXAMPLE

Calculate the hexadecimal equivalent of the decimal number 190.

First, find how many sixteens there are in the number.

$11 \times 16 = 176$

Then find the remainder.

$190 - 176 = 14$

So there are 11 sixteens and 14 remainder. The numbers 11 and 14 are shown as letters in hexadecimal:

11 is B

14 is E.

Put the letters into the position grid.

Sixteens	Ones
B	E

Reading the number from the grid tells you that the decimal number 190 is equivalent to the hexadecimal number BE.

PRACTICE QUESTION

2 Convert these decimal numbers to hexadecimal.

a 60 **b** 125 **c** 79 **d** 220

REMEMBER: To convert a decimal number to hexadecimal: Calculate the number of sixteens in the value. This value goes in the sixteens position. Any remainder goes in the units position. Use hexadecimal digits, including the letters A–F, to represent each value.

1.6 Converting between hexadecimal and binary

Easy conversion

Base 16 (hexadecimal) is used a lot in computer science. That is because it is very easy to convert between binary and hexadecimal. It is much easier than converting either of these number bases into decimal.

There is an exact match-up between the first 16 binary numbers and the 16 hexadecimal digits. You don't need to worry about remainders or subtractions.

Table 1 shows the correspondence. The decimal values are included for reference, but you will not need to look at these when you do binary to hexadecimal conversion.

Table 1 Binary numbers with their equivalent hexadecimal and decimal numbers

Binary	Hexadecimal	Decimal
0000	0	0
0001	1	1
0010	2	2
0011	3	3
0100	4	4
0101	5	5
0110	6	6
0111	7	7
1000	8	8
1001	9	9
1010	A	10
1011	B	11
1100	C	12
1101	D	13
1110	E	14
1111	F	15

There is no need to memorise this table. You can easily work it out by counting up from 0 to 15 in binary and in hexadecimal and writing the numbers next to each other. After doing this a few times, it becomes quite easy to remember the binary and hexadecimal matches.

Binary to hexadecimal

Binary numbers are often shown in groups of eight bits. This is called a byte. You can learn more about bits and bytes in Chapter 3.

A byte is often shown as two groups of four bits. Here is an example:

1011 0100

To convert this number to hexadecimal, write the equivalent hexadecimal digit below each group of four bits.

1011	0100
B	4

So the binary number 1011 0100 is equivalent to hexadecimal number B4.

PRACTICE QUESTION

1 Convert these binary numbers to hexadecimal.

a 0101 0100 **b** 1111 0000
c 0101 1110 **d** 1101 1100

WORKED EXAMPLE

Hexadecimal to binary

Hexadecimal to binary conversion is very similar to the process above.

Convert the hexadecimal number 5C to a binary number.

Write out the digits in the hexadecimal number.

5C

Write the binary equivalent of each digit:

5	C
0101	1100

So the hexadecimal number 5C is equivalent to the binary number 0101 1100.

STRETCH YOURSELF!

You have calculated the hexadecimal equivalents of many binary numbers. To check your working, you can convert both numbers to decimal. They should both come to the same number.

For example, the binary number 1011 0100 is equivalent to the hexadecimal number B4.

Converting 1011 0100 to decimal gives this result:

128	64	32	16	8	4	2	1
1	0	1	1	0	1	0	0

$128 + 32 + 16 + 4 = 180$

Converting B4 to decimal gives this result:

Sixteens	Ones
B	4

Remember, hexadecimal B is equivalent to 11 in decimal.

$11 \times 16 = 176$

$176 + 4 = 180$

So both numbers convert to 180 in decimal.

You can practise this by carrying out this conversion for every hexadecimal/binary pair in this topic.

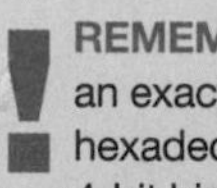

REMEMBER: There is an exact match between hexadecimal digits and 4-bit binary numbers.

To convert between binary and hexadecimal, write the matching hexadecimal number under each group of 4 bits.

To convert between hexadecimal and binary, write the matching four bits under each hexadecimal digit.

2 BINARY CALCULATION

2.1 Binary addition

The rules of addition

The rules of addition are the same in any number system. To add two numbers:

- place the numbers above each other so that the position values are lined up
- add the values of all digits in the same position
- write the result below this column of digits.

To calculate 120 + 58, align the two numbers. The number is shown as far to the right as possible. You may fill empty spaces with 0.

1	2	0
0	5	8
1	7	8

So 120 + 58 = 178

NOTE: If the two values in any position add up to more than nine then the addition will include a number carried to the next position. Addition with carry digits is covered in the next spread.

The rules of binary addition

Binary addition works in exactly the same way as addition in any other number base system. Place the numbers above each other and add up the digits in each column. Because there are only two digits in the binary systems the rules are very simple.

0 + 0 = 0
0 + 1 = 1
1 + 1 = 0 and carry 1.

Carry digits will be covered in Topic 2.2.

WORKED EXAMPLE

Add the two binary numbers: 0101 0001 + 0010 0010.

First, write the numbers in the binary grid. If the numbers have fewer than 8 bits make sure that the number is as far to the right as it can go. Fill in any empty spaces with 0.

0	1	0	1	0	0	0	1
0	0	1	0	0	0	1	0

Now you use the rules of binary addition.

- Where both values are 0; write the result as 0.
- Where one value is 0 and the other is 1; write the result as 1.

0	1	0	1	0	0	0	1
0	0	1	0	0	0	1	0
0	1	1	1	0	0	1	1

This gives the result:

0101 0001 + 0010 0010 = 0111 0011

PRACTICE QUESTION

1 Calculate the results of the following binary additions.

a 0100 1010 + 1000 0100

b 1101 0010 + 0010 1000

c 1001 1010 + 0100 0100

d 0111 0001 + 0000 1100

WORKED EXAMPLE

Calculate the decimal equivalents of the three binary numbers in the following addition sum.

0101 0001 + 0010 0010 = 0111 0011

A good way to check that an addition has been performed correctly is to covert both of the numbers being added and the result into decimal form. The sum should still give the correct answer. This also provides good practice at binary conversion.

There are three numbers. The two that were added together plus the result. The three numbers are placed in the binary position grid.

128	64	32	16	8	4	2	1
0	1	0	1	0	0	0	1
0	0	1	0	0	0	1	0
0	1	1	1	0	0	1	1

Look at each binary number in turn. Add the position values in every place where there is a 1.

$$64 + 16 + 1 = 81$$

$$32 + 2 = 34$$

$$64 + 32 + 16 + 2 + 1 = 115$$

This proves that the binary addition is correct because 81 + 34 = 115.

PRACTICE QUESTIONS

2 Check the results of each binary addition you completed in question 1. Convert all three numbers into decimal. Have you proved that each result is correct? If not, go back and check your working.

3 Calculate the following binary additions and check the results using decimal conversion.

a 1100 1110 + 0001 0000

b 0001 1010 + 0110 0001

c 1001 0010 + 0100 0101

d 0101 0001 + 0000 1000

REMEMBER: The rules of binary addition are:

0 + 0 = 0

0 + 1 = 1

1 + 1 = 0 and carry 1.

2.2 Binary addition with carry bits

Carry digits

When adding numbers in any number system you align the digits by position value. Then you add together the digits with the same position value. Finally, you write the result of this calculation below the digits.

There is one complication. Sometimes the result of the addition is larger than the available digits. For example, in base 10, the result may be larger than nine.

For example, 125 + 58:

1	2	5
0	5	8

In the first column 5 + 8 = 13. This is larger than nine so carry ten into the next column and write in the remaining three:

1	2	3
0	5	8
	1	
		3

In this example the remainder is written above the total row. Sometimes people write the carry digit in a different place but the result is the same. Complete the calculation by adding up the values in each remaining column.

1	2	3
0	5	8
	1	
1	8	3

The completed calculation shows that 125 + 58 = 183.

Carry bits in binary addition

When you add two binary numbers you may need to use carry bits. Binary addition with carry bits can be completed using the following four rules.

$0 + 0 = 0$
$0 + 1 = 1$
$1 + 1 = 0$ and carry 1
1 + 1 + 1 = 1 and carry 1.

These are the rules you learnt in Topic 2.1, with one extra rule.

WORKED EXAMPLE

Calculate the value of 0101 0111 + 0010 0011.

First write the two numbers one above the other in the binary grid.

0	1	0	1	0	1	1	1
0	0	1	0	0	0	1	1

Add an extra row to the grid for carry bits and a final row where the result of the calculation will be shown. Now use the four rules of binary addition to add up each column. The first column is 1 + 1 so put 0 as the column total and carry 1 into the next column.

0	1	0	1	0	1	1	1
0	0	1	0	0	0	1	1
						1	
							0

The second column now includes a carry bit. The addition in this column is 1 + 1 + 1, so put 1 as the column total and carry 1.

0	1	0	1	0	1	1	1
0	0	1	0	0	0	1	1
					1	1	
						1	0

Now complete the binary addition in each column. Use carry bits where needed. Here is the completed calculation:

0	1	0	1	0	1	1	1
0	0	1	0	0	0	1	1
				1	1	1	
0	1	1	1	1	0	1	0

The completed calculation shows us that:

0101 0111 + 0010 0011 = 0111 1010

PRACTICE QUESTIONS

1 Calculate the following binary additions.

a 0011 1101 + 0000 1101

b 1001 1001 + 0001 1111

c 1001 1010 + 0100 1110

d 0111 0001 + 0001 1111

2 Check the addition sums you completed in question 1.

Convert each number and the result into decimal. If the numbers do not add up, find and correct the error.

3 Calculate the following binary additions. The numbers are not split into groups of 4 bits.

a 101101 + 1011111

b 1100101 + 110001

c 10010101 + 1100

d 1011101 + 1010011

Use decimal conversion to check your results.

> **REMEMBER:** There is an extra rule of binary addition when working with carry bits.
>
> 1 + 1 + 1 = 1 and carry 1.

2.3 Add three binary numbers

Add three binary numbers

Sometimes you may need to add more than two binary numbers together. In many cases you can complete this type of addition using the rules of binary addition that you have already learnt.

$0 + 0 = 0$
$0 + 1 = 1$
$1 + 1 = 0$ and carry 1
$1 + 1 + 1 = 1$ and carry 1.

More complicated binary additions, involving more than 3 bits, are beyond the scope of this book.

WORKED EXAMPLE

Calculate the result of adding the three binary numbers 0101 0111 + 0010 0011 + 0000 1100.

First, align the three numbers in the binary grid.

0	1	0	1	0	1	1	1
0	0	1	0	0	0	1	1
0	0	0	0	1	1	0	0

Then use the rules of binary addition to add up each column. Show carry bits where needed.

The first column is 1 + 1. Write 0 and carry 1. With the carry bit, the second column is 1 + 1 + 1. Write 1 and carry 1.

0	1	0	1	0	1	1	1
0	0	1	0	0	0	1	1
0	0	0	0	1	1	0	0
					1	1	
						1	0

Continue to add up each column using the rules of binary addition. Here is the completed grid:

0	1	0	1	0	1	1	1
0	0	1	0	0	0	1	1
0	0	0	0	1	1	0	0
	1	1	1	1	1	1	
1	0	0	0	0	1	1	0

This gives the result:

0101 0111 + 0010 0011 + 0000 1100 = 1000 0110

PRACTICE QUESTIONS

1 Calculate the results of the following binary additions.

a 0011 1101 + 0000 1101 + 0100 0000

b 1001 1001 + 0001 1111 + 0010 0101

c 1001 1010 + 0100 1110 + 0001 0001

d 0111 0001 + 0001 1111 + 0010 0011

2 Check your answers to the addition sums in question 1 by converting each number and the result into decimal. If the numbers do not add up, find and correct the error.

3 Calculate the results of the following binary additions.

a 0101 1101 + 0010 1001 + 0010 0011

b 1001 0011 + 0100 1001 + 0001 0101

c 0001 1101 + 1000 0011 + 0100 1100

d 0100 1100 + 0010 1010 + 0011 0001

4 Check your answers to the addition sums in question 3 by converting each number and the result into decimal. If the numbers do not add up, find and correct the error.

STRETCH YOURSELF!

Larger carry values

In decimal numbers 1 + 1 + 1 = 3. In binary we write this as 1 and carry 1.
The carry bit represents 2.

When adding three binary numbers you can occasionally end up needing to carry a value greater than two. For example:

Add 0011 + 0010 + 0111

When the numbers are placed in a binary grid. The first column is 1 + 1. The total is 0 and carry 1.

0	0	1	1
0	0	1	0
0	1	1	1
		1	
			1

The second column now contains 1 + 1 + 1 + 1. The answer is four.

To show this quantity in binary we write 0 and carry 10 (10 is four in binary). That means we must add the carry bit two columns to the left, as shown in the next table.

0	0	1	1
0	0	1	0
0	1	1	1
1	0	1	
		0	0

Completing the addition gives 1100.

Converting these numbers to decimal gives you the sum 3 + 2 + 7 = 12, which is correct.

2.4 Overflow errors

Numbers represented by one byte

Binary numbers are created from bits – the binary digits 1 and 0. Bits are typically collected in groups of eight. A group of 8 bits is called a byte. The memory inside the computer is composed of bits organised into bytes.

The largest number that can be represented in 1 byte is 1111 1111. To find its decimal value place it into the binary grid.

128	64	32	16	8	4	2	1
1	1	1	1	1	1	1	1

The value of this number is:

$$128 + 64 + 32 + 16 + 8 + 4 + 2 + 1 = 255$$

255 is the largest number that can be represented with one byte of data.

Numbers larger than one byte

A number that is larger than 255 cannot be represented using a single byte of data. A computer will store numbers beyond 255 by linking several bytes together.

When you add together two binary numbers you may get a result that is larger than 255. If the result is to be shown using a single byte then this will produce an error. The name for this is an overflow error. You should indicate that you recognise that there has been an error, and the type of error that has occurred.

This happens when a carry bit results from adding the values in the left-hand column. There is no 'next position' to hold the carry bit because you are already working on the left-most position of the byte.

WORKED EXAMPLE

Calculate 0111 0111 + 1010 0011 and give your answer as one byte.

Place the numbers in the binary grid, begin adding the digits in each position, and use carry bits where necessary.

Here is the calculation almost completed. Only the final column needs to be completed.

0	1	1	1	0	1	1	1
1	0	1	0	0	0	1	1
1	1			1	1	1	
	0	0	1	1	0	1	0

The final column has the calculation 1 + 1. The rules of binary arithmetic say 1 + 1 = 0 and carry 1 – but there is no further column to hold the carry bit.

	0	1	1	1	0	1	1	1
	1	0	1	0	0	0	1	1
1?	1	1			1	1	1	
	0	0	0	1	1	0	1	0

You should therefore write the result as:

0111 0111 + 1010 0011 = 0001 1010 **with an overflow error**

Note that the answer could be shown correctly with 9 bits. You should only state that there is an overflow error if you are asked to give the answer as a single byte.

PRACTICE QUESTION

1 Calculate the results of the following binary additions. Give your answer as one byte and indicate whether there is an overflow error.

a 0110 0100 + 1100 1000

b 0111 1000 + 0111 0101

c 1001 0110 + 0110 0011

d 1011 0100 + 0100 1101

WORKED EXAMPLE

Check by converting to decimal

You can check if there is an overflow error by converting the calculation into decimal.

In the example above you carried out the binary calculation:

0111 0111 + 1010 0011

To find their decimal values, place these numbers in the binary grid.

128	64	32	16	8	4	2	1
0	1	1	1	0	1	1	1
1	0	1	0	0	0	1	1

The first number is:

64 + 32 + 16 + 4 + 2 + 1 = 119

The second number is:

128 + 32 + 2 + 1 = 163

Adding these two together gives:

119 + 163 = 282

The result is larger than 255 so it will produce an overflow error.

PRACTICE QUESTION 2

2 Convert each number in question 1 into decimal to check for overflow errors.

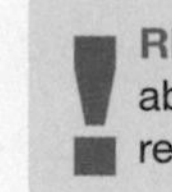

REMEMBER: Numbers above 255 cannot be represented using 8 bits.

2.5 Binary multiplication

Multiply by two

In a binary number the position on the far right has a value of 1. Each position value, moving to the left, is two times larger. If you shift all the bits to the left, the value of the number is doubled.

So to multiply a binary number by two, you shift every bit one place to the left.

WORKED EXAMPLE

Multiply the binary number 0001 0010 by two.

Before the shift:

128	64	32	16	8	4	2	1
0	0	0	1	0	0	1	0

This number has the value 16 + 2 = 18.

To multiply the number by two, shift every bit one position to the left.

128	64	32	16	8	4	2	1
0	0	1	0	0	1	0	0

The result is 0010 0100.

The decimal value of this number is 32 + 4 = 36.

You can check the result by converting the original number to a decimal and multiplying by two:

$18 \times 2 = 36$

PRACTICE QUESTION

1 Multiply the following binary numbers by two.

a 0110 0100 **b** 0001 1000 **c** 0001 0110 **d** 0111 0100

Convert to decimal to check your result.

Powers of two

A power of two means a number produced by multiplying two by itself.

For example, 16 is $2 \times 2 \times 2 \times 2$. You can also write this as 2^4 or 'two to the power four'. This expression means that two is multiplied by itself four times.

In mathematics a number to the power 0 has the value 1. That means $2^0 = 1$ is also a power of two.

Table 1 Powers of two expressed as decimal and binary numbers

Power of two	As a decimal number	As a binary number
2^0	1	1
2^1	2	10
2^2	4	100
2^3	8	1000
2^4	16	10000
2^5	32	100000
2^6	64	1000000

The powers of two are the same as the binary position values. The number of zeros in the binary number is the same as the power of two. For example, 2^6 has six zeros.

You can use binary shift to multiply a binary number by any power of two. Select what power of two you want. Then shift bits to the left by that number of places.

How multiplication questions are shown

You may see a binary multiplication in this form:

multiply the binary number 0001 0010 by four.

Or in this form:

multiply the binary numbers 0001 0010 and 100.

The second value – the multiplier – may be shown as a decimal number or as a binary number. These both represent the same calculation. In either case you must shift the bits two places to the left because four is two to the power two.

WORKED EXAMPLE

Multiply the binary number 0001 0010 by eight.

Place the binary number in the grid:

128	64	32	16	8	4	2	1
0	0	0	1	0	0	1	0

Eight is 2^3 (two to the power three). Multiplying a binary number by eight means you must shift every bit three places to the left:

128	64	32	16	8	4	2	1
1	0	0	1	0	0	0	0

The decimal value of the first number is 16 + 2 = 18.

The decimal value of the second number is 128 + 16 = 144.

$18 \times 8 = 144$

The binary shift has multiplied the number by eight.

PRACTICE QUESTION

2 Use binary shift to multiply:

- **a** the binary numbers 0001 0100 and 1000
- **b** the binary numbers 0001 1000 and 100
- **c** 0001 0110 by the decimal number four
- **d** 0000 0111 by the decimal number 16.

Convert each answer to decimal to check your results.

REMEMBER: Shifting the bits of a binary number one place to the left multiplies the number by two.

You can multiply by any power of two using this method.

2.6 Binary division

Divide by two

Shifting binary digits to the left makes their values larger. Shifting bits to the right makes their values smaller.

- Left-shifts are used to multiply binary numbers by two or by powers of two.
- Right-shifts are used to divide binary numbers by two or powers of two.

It is easy to remember: positions on the right of the binary grid have a lower value. So shifting bits in that direction gives the number a smaller value.

WORKED EXAMPLE

Divide the binary number 0001 0010 by two.

Before the shift:

128	64	32	16	8	4	2	1
0	0	0	1	0	0	1	0

This number has the value 16 + 2 = 18

To divide the number by two shift every bit one place to the right.

128	64	32	16	8	4	2	1
0	0	0	0	1	0	0	1

The decimal value of this number is 8 + 1 = 9.

$$\frac{18}{2} = 9$$

The binary shift has divided the number by two.

PRACTICE QUESTION

1 Divide the following binary numbers by two.

a 0110 0100 **b** 0001 1000

c 0001 0110 **d** 0111 0100

Convert each number to decimal to check your results.

How division questions are shown

You may see a binary division in this form:

divide the binary number 0010 0100 by four.

or in this form:

divide the binary number 0010 0100 by the binary number 100.

The second value – the divisor – may be shown as a decimal number or as a binary number. These both represent the same calculation. In either case you must shift the bits two places to the right because four is two to the power two.

WORKED EXAMPLE

Divide the binary number 0111 0000 by the decimal number 16.

Place the binary number in the grid.

128	64	32	16	8	4	2	1
0	1	1	1	0	0	0	0

The decimal value of the binary number is 64 + 32 + 16 = 112.

16 is 2^4 (two to the power four). Dividing a binary number by 16 means you must shift every bit four places to the right. This is shown in the next table.

128	64	32	16	8	4	2	1
0	0	0	0	0	1	1	1

The decimal value of the second number is 4 + 2 + 1 = 7.

$$\frac{112}{16} = 7$$

The binary shift has divided the number by 16.

NOTE: Sometimes a bit is shifted so far to the right that there is no more space in the binary grid. The bit 'falls off' the end of the grid on the right. This is called an underflow error.

PRACTICE QUESTION

2 Use binary shift to divide:

a the binary number 0001 1100 by the binary number 10

b the binary number 1010 0000 by the binary number 0001 00000

c the binary number 1101 0100 by the decimal number four

d the binary number 0101 1000 by the decimal number eight.

Convert your answers to decimal to check them.

Conclusion

Each bit in a binary number gets its value from its position. Binary shift means moving all the bits in a number. All the bits move the same number of places to the right or the left.

- A left-shift moves all the bits in a number to the left. Moving a bit to the left gives it a higher value. Left shift has the effect of multiplying the number. The number is multiplied by a power of two.
- A right-shift moves all the bits to the right. Moving a bit to the right gives it a lower value. Right shift has the effect of dividing the number. The number is divided by a power of two.

The number of shifts to the right or left tells you the power of two that is used.

REMEMBER: Shifting the bits of a binary number one place to the left divides the number by two.

You can divide by any power of two using this method.

3 DATA REPRESENTATION

3.1 Digital data

What does 'digital' mean?

All data inside the computer is stored in digital form. The data is held using microscopic electronic circuits. Each circuit can either hold an electrical charge or no charge. These two states (on and off) are equivalent to the 1s and 0s of binary numbers.

The computer does not literally hold millions of little 1s and 0s. It holds on/off electrical signals which represent the numbers that are being stored.

The electrical signals inside the computer are constantly changing as the computer carries out logical and mathematical operations.

Data and instructions

Everything that is held or processed by a computer must be turned into electronic on/off signals.

- All data content, including letters of the alphabet, pictures and sounds, must be turned into digital format.
- The instructions that the computer follows are also held in digital form. The number code that is used to store computer instructions is called *machine code.*

Both data and instructions are held in memory and stored in files. A file is an organised collection or digital data or machine code instructions. The file is given a name. The size of the file tells you how much storage space it uses up.

Figure 1 shows some of the files stored on a computer system. File size is shown in kB (kilobytes). A kilobyte is 1000 bytes.

Name	Date modified	Type	Size
blue_banner_light.png	30/04/2018	PNG File	20 kB
Schedule.csv	25/04/2018	CSV File	25 kB
codes.txt	13/10/2017	Text Document	1 kB
To do list.doc	11/07/2018	DOC File	12 kB

Figure 1 File sizes displayed in a computer's file system

PRACTICE QUESTION

1 a Identify four files on your computer. In each case give the name of the file and the size of the file.

b State the units used to show file size.

c Give the largest type of file and state what is stored in this file. If it is data, give the type of data.

Bits and bytes

The size of computer memory is measured in bits and bytes.

- *Bit* – a single electronic circuit can represent a 1 or a 0. One circuit can hold 1 bit of information.

- *Byte* – inside the computer, electronic circuits are grouped into eights. A group of 8 bits is called a byte.

It is easy to convert between bits and bytes.

- If you know the number of bits in a file, divide by eight to give the number of bytes.
- If you know the number of bytes in a file, multiply by eight to give the number of bits.

It is important to distinguish between bits and bytes. Bits are sometimes shown with a lower case 'b'. Bytes are sometimes shown with an upper case 'B'. For example:

- 64 b means 64 bits
- 64 B means 64 bytes.

In this book the full word 'bit' or 'byte' is used to make it clear.

PRACTICE QUESTION

2 Complete the table with the missing values.

File name	Size in bits	Size in bytes
User manual.txt		28 750
Weather report.doc	190 000	
Merge sort.py		3000
Holiday.html	35 000	

Nibble

Computer scientists sometimes talk about an intermediate unit called a nibble. A nibble is a group of 4 bits. Another way to say it is that a nibble is half a byte.

Converting bits or bytes to nibbles is an easy calculation.

- If you know the number of bits in a file, divide by four to give the number of nibbles.
- If you know the number of bytes in a file, multiply by two to give the number of nibbles.

The name nibble is probably based on the fact that taking half a bite out of a piece of food might be called a nibble.

PRACTICE QUESTION

3 Add an extra column to the table you made for practice question 2. Give the number of nibbles in each file.

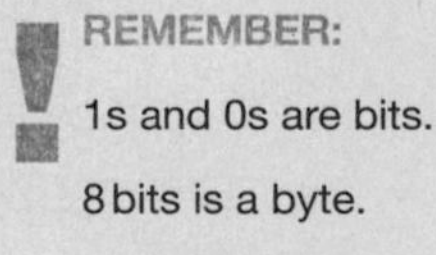

REMEMBER:

1s and 0s are bits.

8 bits is a byte.

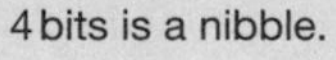

4 bits is a nibble.

3.2 Units of digital data

Large computer files

Modern computer files can be quite large. They can contain millions or even billions of bytes. For example, a computer file storing a digital movie might contain 4 000 000 000 bytes. It is difficult to read large numbers like this and easy to make mistakes when you write them down.

Large numbers of bytes are usually represented using other units, such as kilobyte (kB) and megabyte (MB). The number of bytes in each unit is shown in the next table.

Unit	Abbreviation	Meaning	Total number of bytes
byte	B	8 bits	1
kilobyte	kB	1000 bytes	1 000 (a thousand)
megabyte	MB	1000 kilobytes	1 000 000 (a million)
gigabyte	GB	1000 megabytes	1 000 000 000 (a billion)
terabyte	TB	1000 gigabytes	1 000 000 000 000 (a trillion)
petabyte	PB	1000 terabytes	1 000 000 000 000 000 (a quadrillion)

PRACTICE QUESTION

1 Complete the following table showing file sizes in words and then in number form with unit abbreviations.

File size in words	In numbers
one hundred and fifty kilobytes	150 kB
half a megabyte	
	5 PB
four terabytes	

Converting between units

Converting between units generally means multiplying or dividing by 1000. For example, a megabyte is 1000 kilobytes.

- To convert megabytes to kilobytes you multiply by 1000.
- To convert kilobytes to megabytes you divide by 1000.

Remember that when you convert a value to smaller units the number increases. But if you convert to larger units the number of units decreases.

PRACTICE QUESTION

2 Convert:

a 34 300 bytes to kilobytes

b 1 700 000 bytes to megabytes

c 0.45 gigabytes to megabytes

d 0.003 terabytes to gigabytes

e 50 000 0000 megabytes to petabytes

f 0.0001 megabytes to bytes

Which unit to choose

The purpose of these large units is to make values easier to read, write, and understand. Therefore, use a unit that lets you express a value in the simplest way, using the smallest number of digits.

For example, these expressions all represent the same file size:

1 200 000 bytes
1.2 MB
0.0012 GB.

Expressing the value in bytes uses seven digits, expressing the value in megabytes uses two digits, and expressing the value in gigabytes uses five digits.

The value expressed in megabytes (1.2 MB) uses the fewest digits. It is easiest to read and understand making it the best unit to choose.

WORKED EXAMPLE

Express 1 300 000 000 bytes using the most suitable unit.

The quantity is very large. It will be best to convert it to a larger unit. To convert bytes into kilobytes divide by 1000.

1 300 000 000 bytes = 13 000 000 kB

However, this number is still very large. Divide kilobytes by 1000 to convert to megabytes.

1 300 000 kB = 1300 MB

The number is still large so you can divide by 1000 again. This converts megabytes to gigabytes.

1300 MB = 1.3 GB

At this point the quantity is expressed in its simplest form. Further division would increase the number of digits.

PRACTICE QUESTION

3 Convert the following file sizes to the most suitable unit of measurement. The aim is to make the expression simple to read and understand.

- **a** 0.0004 PB
- **b** 1 750 000 MB
- **c** 7 000 000 000 kB
- **d** 0.000 001 TB

REMEMBER: If the number of bytes is large express it using larger units.

The units are kilobytes, megabytes, gigabytes, terabytes, and petabytes.

Each unit is 1000 times bigger than the previous unit.

NOTE: Sometimes 'kilobyte' is used to mean a group of 1024 bytes, and 'megabyte' to mean a group of 1024 kilobytes. This is because 1024 is a power of 2 (2^{10}).

You should check whether your exam board expects you to treat a kilobyte as 1000 or 1024 bytes. For Edexcel, a kilobyte is defined as 1024 bytes in questions about data storage.

3.3 How much storage?

The importance of file size

To hold a file, a computer must have enough storage space. Storage space and file size are expressed in the digital units you have learnt, such as bytes, kilobytes, and megabytes.

Before storing digital content you must know:

- how much storage space is available
- the size of the file or files you want to store.

The size of the files must be smaller than the storage space.

Understanding file size is also important when data is transmitted, for example, over the Internet. A large file will be much slower to transmit.

Work out total file size

To work out total storage requirements find the size of each file and add the values together. All file sizes must be expressed using the same digital units. The result is also expressed using the same unit.

PRACTICE QUESTION

1 Add together the following file sizes to give the total storage requirement.

a 40MB + 30MB

b 34GB + 156GB + 43GB + 216GB

c 78kB + 123kB + 99kB

d 2TB + 9TB + 15TB

Simplifying the expression

If the total file size is a very large value; simplify the expression using the methods you learnt in Topic 3.2. Remember, the goal is to reduce the number of digits and make the number easier to read.

WORKED EXAMPLE

A student has two video files that are 250 MB and 950 MB. Calculate the total storage space she needs to store these two files. Choose a suitable unit to express your answer.

The total storage requirement is calculated by adding together the two file sizes.

250MB + 950MB = 1200MB

Both file sizes were expressed in megabytes. The result is expressed in megabytes.
The result is more than 1000 megabytes so it is appropriate to simplify the answer. The total file size can be converted to gigabytes:

1200MB = 1.2GB

PRACTICE QUESTION

2 Calculate the total storage requirement of the following pairs of file sizes. Give the result using a suitable unit.

a 452 MB + 690 MB
b 712 000 bytes + 350 000 bytes
c 94 000 kB + 800 000 kB
d 20 TB + 90 TB

Use common units

To add file sizes together all values must be expressed using the same units. If values are expressed using different units then convert the values so that they all use the same units. Then you can add the file sizes as before.

WORKED EXAMPLE

A web designer has 3.5 MB of storage space available on a server. Determine if there is space for the following three files.

File name	File size
a.bmp	2.5 MB
b.png	340 kB
c.jpg	0.09 MB

These values are expressed in two different units – megabytes and kilobytes. Before adding them together they must all be converted to the same unit. You can choose any unit, but perhaps best to choose megabytes as that means you only need to convert one value.
First convert 340 kB to MB. To do this divide by 1000:

$$\frac{340}{1000} = 0.34\text{ MB}$$

Then add all the values together. The result is a value in megabytes:

$$2.5 + 0.34 + 0.09 = 2.93\text{ MB}.$$

2.93 MB is less than the available space of 3.5 MB, so the files will fit in the storage.

PRACTICE QUESTION

3 Calculate the total storage requirement for each of the following pairs of file sizes. Give the result using a suitable unit.

a 4 MB + 0.0056 GB
b 12 000 bytes + 350 kB
c 94 000 MB + 0.092 GB
d 2 TB + 50 000 MB

REMEMBER: Change file sizes to the same units before adding them together.

Express the result using the most suitable unit.

3.4 Representing characters with ASCII

Digital codes

Data inside the computer must be represented using on/off signals – used to represent the 1s and 0s of binary numbers. All data inside the computer must be stored in this form, using number codes.

ASCII

ASCII is a number code system. Every keyboard character has a number code. ASCII characters include letters of the alphabet, digits, punctuation marks and other symbols, and control signals such as 'Escape' or 'Tab'.

When you press a key on your keyboard a signal is sent from the keyboard to the computer. The computer recognises which key you have pressed. It stores the ASCII code number that matches that key.

PRACTICE QUESTION

1 The core ASCII character set has 128 number codes (from 0 to 127).

  a Using your knowledge of binary numbers, suggest how many bits are needed to store the code for one character.

  b The ASCII code for 'A' is number 65. Give this as a binary number.

Character set

A character set is any list of characters. Each character is given a code number. The codes are usually allocated in alphabetical order. That means if you know the code for A and B you can work out the code for C. Here is part of the ASCII character set. It shows the lower-case letters.

ASCII character	ASCII code in decimal	ASCII code in binary
a	97	011036 0001
b	98	0110 0010
c	99	0110 0011

PRACTICE QUESTION

2 Draw the ASCII table showing all the characters from 'a' to 'z' and for each character complete:

  a the column showing ASCII codes in decimal

  b the column showing the binary version of the ASCII code.

NOTE: ASCII stands for American Standard Code for Information Interchange. ASCII was introduced in 1963. In 1981 IBM adopted ASCII for its first personal computer. From then on ASCII grew in popularity to overtake all other character coding systems.

Sequential numbering

You do not have to memorise all the ASCII character codes. Because ASCII codes are given in sequence, you can work out the code for a letter of the alphabet if you are given any other alphabet code. Just count backwards or forwards from the one you know.

WORKED EXAMPLE

The ASCII code for A is 65. Give the ASCII code for G in decimal and binary.

Character	ASCII code
A	65
B	66
C	67
D	68
E	69
F	70
G	71

We know that ASCII codes are sequential so we count on from A to G.

This shows that the ASCII code for 'G' is 71 (in decimal).

Using the binary grid we can convert decimal 71 into binary.

$64 + 4 + 2 + 1 = 71$

128	64	32	16	8	4	2	1
0	1	0	0	0	1	1	1

So G is represented in ASCII by the binary number 01000111.

PRACTICE QUESTION

3 The decimal code for the 'space' character between words is 32.

a Convert the following message into a series of ASCII codes. Remember to use the correct codes for upper and lower case letters.

'This website uses cookies'

b Translate the following series of character codes into a text message.

112 97 115 115 119 111 114 100 32 105
115 32 117 110 105 99 111 114 110

Digits

Digits are keyboard characters too. When you type a digit on your keyboard the computer will store the data first of all using an ASCII value.

Notice that the ASCII codes do not match the number values of these digits. ASCII is a code that records which characters you have typed on the keyboard. Later, if the computer needs to perform a calculation, these digits may be converted into their actual number values.

The ASCII codes are sequential so you can work out the codes for all the other digits by counting upwards.

Table 2 ASCII number codes for the digits 0, 1, and 2 on the keyboard

Digit	ASCII code
0	48
1	49
2	50

PRACTICE QUESTION

3 a Draw the ASCII table showing the digits from 0 to 9 together with their equivalent ASCII codes.

b Add an extra column to the table to show the binary equivalent of the ASCII code.

REMEMBER: Every digit from 0 to 9 has an ASCII code. The ASCII code can be written in decimal or binary.

This code is NOT the number value of the digit. It is just a code to let the computer identify the keyboard character.

ASCII codes are sequential.

3.5 Unicode and extended ASCII

7-bit ASCII

The standard ASCII character set has 128 different character codes. To represent the numbers from 0 to 127 requires up to 7 bits. Computer memory is organised into groups of 8 bits called bytes. 7-bit ASCII uses the first 7 bits for the number code. The eighth bit is used for error checking and is kept free in this system.

PRACTICE QUESTION

One character in 7-bit ASCII takes up 1 byte of computer memory.

1 An essay has 2583 words, an average word length of five characters, a space character after every word except the last one, and a punctuation mark after every ninth word.

Calculate how many bytes of memory are needed to store the report in 7-bit ASCII. Give the answer in suitable binary units.

Extended ASCII

7-bit ASCII only has codes for 128 characters – not enough to represent all the different characters that a user might want to enter into a computer. For this reason, several new versions of ASCII were developed with longer codes and larger character sets. The general term for these versions is 'extended ASCII'.

- Most extended ASCII character sets use all 8 bits of the byte to make a code. 8 bits allow a bigger range of numbers than 7 bits.
- A few versions of extended ASCII use more than 1 byte to store the code. This allows a much bigger set of code numbers.

The different versions of extended ASCII used different character sets. That meant a particular code – for example 200 – could stand for different characters in different versions of ASCII. Computers using different number codes could cause problems and confusion in communication.

PRACTICE QUESTION

2 7-bit ASCII has 128 different number codes. State how many number codes there are in 8-bit ASCII.

Unicode

To overcome this confusion a new coding system was established in 1987. The new system was called Unicode. Unicode aims to provide a code for every character in every alphabet world-wide. Unicode also has codes for characters such as mathematical and scientific symbols and 'emojis' (picture symbols).

The first 128 characters in the Unicode set are exactly the same as 7-bit ASCII. That means any text coded in ASCII can be read by a computer that understands Unicode.

NOTE: Unicode is controlled by a central authority called the Unicode Foundation. The Unicode Foundation decides which character code goes with which character. This means there is much less chance of confusion.

PRACTICE QUESTION

3 Convert the following message from these Unicode values to text.

84 104 101 32 102 105 114 115 116 32 49 50 55 32 99 104 97 114 97 99 116 101 114 32 99 111 100 101 115 32 97 114 101 32 116 104 101 32 115 97 109 101 32 97 115 32 65 83 67 73 73

How many code numbers are there in Unicode?

Bigger code numbers take more storage space (see Table 1).

- 1 byte of data can store 256 different codes.
- 2 bytes of data can store 256 × 256 = 65 536 different codes.

Standard Unicode can use up to 4 bytes per character. This gives a lot of codes.

Table 1 Unicode storage bytes

Number of bytes	Number of codes	Biggest code value
1	256	256
2	256 × 256	65 536
3	256 × 256 × 256	16 777 216
4		

PRACTICE QUESTION

4 Complete Table 1 to show the biggest code number that can be stored using four bytes of data.

REMEMBER: ASCII is a family of character sets. This includes 7-bit ASCII, the various types of extended ASCII, and Unicode.

The codes from 0 to 127 are the same in every version. So the code 65 always stands for 'A' in ASCII or Unicode.

Calculate file size

It is difficult to calculate file size in Unicode. That is because Unicode is often stored in a flexible way. One text character is stored using a single character code, but:

- some Unicode values are small numbers
- some Unicode values are very large numbers.

In most Unicode systems the computer chooses how many bytes of storage to use for each character. If the character has a small code number the computer uses just 1 byte. If the character code is a large number the computer might use up to 4 bytes to store that character code.

SUMMARY QUESTIONS FOR CHAPTERS 1–3

1 Convert the following binary numbers to decimal.

a 1111 0000

b 1011 1101

c 1110 1110

d 1000 1000

2 Convert the following decimal numbers to binary.

a 61

b 28

c 172

d 203

3 Convert the following hexadecimal numbers to decimal.

a 29

b 14

c A6

d BD

4 Convert the following decimal numbers to hexadecimal.

a 165

b 58

c 181

d 47

5 Convert the following binary numbers to hexadecimal.

a 1011 0011

b 0100 1110

c 0001 1111

d 0101 0001

6 Convert the following hexadecimal numbers to binary.

a F5

b 9D

c DA

d 88

7 Complete the following binary additions.

a 1011 0011 + 0100 1000

b 1011 1010 + 0001 0101

c 1011 1010 + 0011 1010

d 0101 1100 + 0010 0101 + 0101 0001

8 a 0111 0110 + 0100 1110 + 0001 1001

b 0100 1100 + 1000 1001 + 0010 1001

c 0011 1111 + 0100 1101 + 0001 0001

d 1001 1101 + 0101 0101 + 0000 1001

9 Convert the binary sums in Q8 into denary to check your answers.

10 Multiply the binary number by the decimal value.

a 0101 1100 × 2

b 0001 1111 × 8

11 Multiply the two binary numbers.

a 0010 1011 × 0100

b 0100 1110 × 0010

12 Divide the binary number by the decimal value.

a 0110 1100 divided by 2

b 1100 1000 divided by 8

13 Convert the following file sizes to the simplest form using appropriate units.

a 6 600 000 bytes

b 0.0001 MB

c 15 000 GB

d 0.000 009 TB

14 Complete the table to show the missing ASCII codes in decimal and binary.

Letter	Decimal ASCII code	Binary ASCII code
R	82	
S	83	
T		0101 0100
U		
V		

4 DIGITAL IMAGES

4.1 Bitmap images

Pixels

Computer images are made up of pixels. Pixel is short for picture element. A pixel is the smallest part of an image that can be edited. That is, the computer stores information about each pixel and can change an image by as little as one pixel.

When an image is displayed on the screen, the pixels may be the small points of light that make up the image. When an image is printed out, a pixel may be one spot of the ink or toner that makes the printed image.

All data inside the computer is represented in number form. Inside the computer each pixel is represented by a number value. There are different ways of representing pixels in number form. You will learn more in this section.

Pixel grid

On the computer screen pixels are very small and there may be thousands or millions of pixels in a large image. Whatever the size of the image or the number of pixels, they are arranged into a grid. Here is a small example:

- This grid is ten pixels wide and ten pixels high.
- It contains 10 × 10 = 100 pixels.

Monochrome image

Monochrome means one colour. In fact, monochrome images are usually made of one colour plus white. Most monochrome images are black and white. The computer can record a monochrome image using a series of bits. The two values of the bits – 1 and 0 – stand for the two colours.

WORKED EXAMPLE

Here is a monochrome image made using the pixel grid.

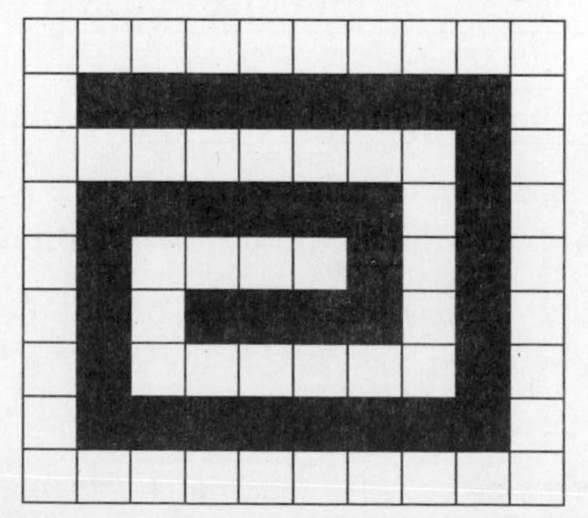

To record this image the computer must turn it into a series of bits.
The computer will assign the two colours to the two bit values. For example:

1 = white
0 = black

We can show these two values in the pixel grid.

1	1	1	1	1	1	1	1	1	1
1	0	0	0	0	0	0	0	0	1
1	1	1	1	1	1	1	1	0	1
1	0	0	0	0	0	0	1	0	1
1	0	1	1	1	1	0	1	0	1
1	0	1	0	0	0	0	1	0	1
1	0	1	1	1	1	1	1	0	1
1	0	0	0	0	0	0	0	0	1
1	1	1	1	1	1	1	1	1	1

Now take these values out and make a list of all the 1s and 0s. They must remain in the same order as they appear in the grid.

1111111111
1000000001
1111111101
1000000101
1011110101
1010000101
1011111101
1000000001
1111111111

Finally, keeping the order of bits the same, make them into one long series – sometimes called a bitstream.

111111111110000000011111111101100000010110111101011010000101101111110110
000000011111111111

This bitstream will let the computer store the image. The only other piece of information it needs is the width and height of thc pixel grid.

PRACTICE QUESTION

1 Convert the monochrome bitmap image (shown on the right) into a bitstream.

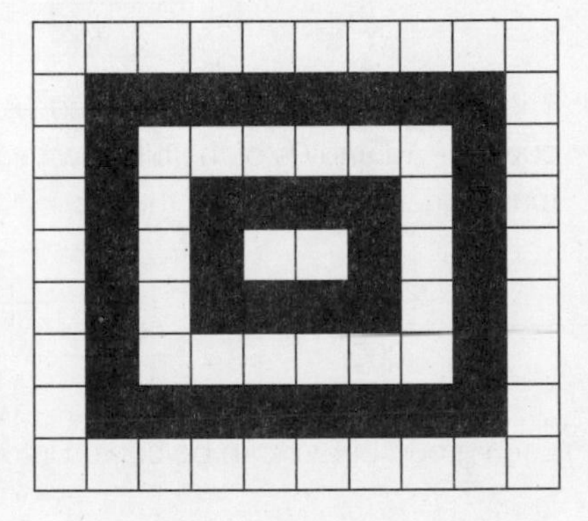

How many bits are needed?

In a monochrome image every pixel is stored using a single bit. The size of the pixel grid tells you how many pixels there are in the image. That tells you how many bits are needed.

A grid that is 100 pixels wide and 200 pixels high will have $100 \times 200 = 20\,000$ pixels.

The image will be stored using 20 000 bits.

The size – also called the resolution – is given as width × height, in that order.

PRACTICE QUESTION

2 A graphic designer created some icons for a computer game. Each icon was created in monochrome on a bitmap grid. State how many bits it would take to store each design.

a EXIT SYMBOL: a grid 15 pixels wide and 20 pixels high

b TREASURE: a grid 12 pixels wide and 16 pixels high

c WARNING: a grid 26 pixels wide and 22 pixels high

d WEAPON CACHE: a grid 25 pixels wide and 32 pixels high

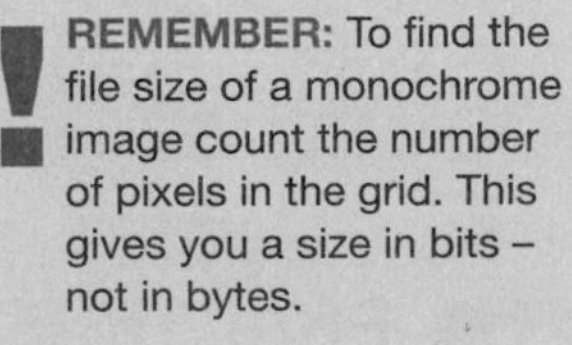

4.2 Turn a bitstream into an image

Make an image

A monochrome image can be created from the 1s and 0s of a bitstream.

- Put the 1s and 0s into a pixel grid.
- Replace 1s with white and 0s with black.

The only other information needed is the width and height of the pixel grid.

WORKED EXAMPLE

Interpret the following bit stream as a bitmap image in a 4 × 4 grid.

1101110100001101

A 4 × 4 grid looks like this:

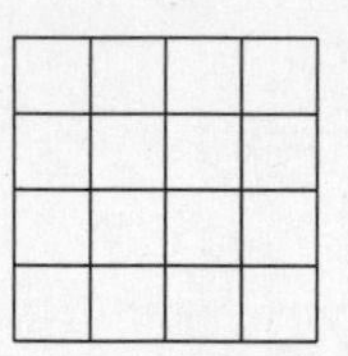

There are 16 pixels in the grid, and 16 bits in the bitstream. Starting in the top left hand corner, copy the 1s and 0s of the bitstream into the grid. It may be helpful to split them into groups of four to match the rows of the grid.

1101
1101
0000
1101

Each group of four can be copied into one row of the grid.

1	1	0	1
1	1	0	1
0	0	0	0
1	1	0	1

The 1s become white pixels and the 0s become black pixels.

1	1	0	1
1	1	0	1
0	0	0	0
1	1	0	1

This is the completed image.

PRACTICE QUESTION

1 Convert the following bitstreams into bitmap images.

a 1100 1100 0011 0011 in a 4 × 4 grid

b 10101 00000 10101 00000 10101 in a 5 × 5 grid

c 110011 101101 111101 110011 110111 111111 110111 in a 6 × 7 grid

WORKED EXAMPLE

Convert the following bitstream to a bitmap image in a 6 × 8 grid.

111111100111101111011100011101011100001111111

The grid is described as 6 × 8. This means the width is 6 and the height is 8.

Add the pixels to the grid starting in the top left hand corner. Then shade in the pixels marked with a 0.

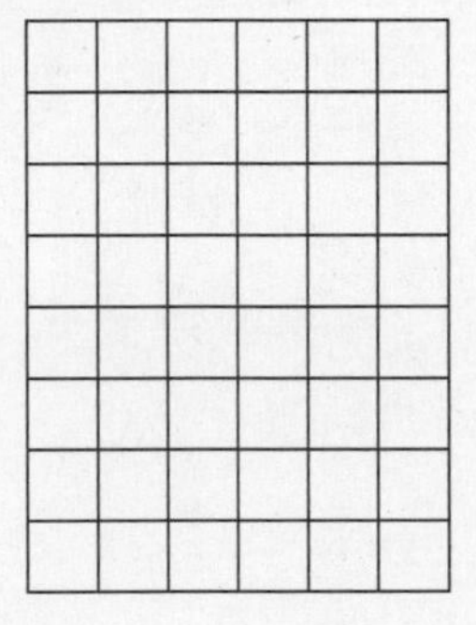

1	1	1	1	1	1
1	1	0	0	1	1
1	1	1	0	1	1
1	1	1	0	1	1
1	0	0	0	1	1
1	0	1	0	1	1
1	0	0	0	0	1
1	1	1	1	1	1

1	1	1	1	1	1
1	1	0	0	1	1
1	1	1	0	1	1
1	1	1	0	1	1
1	0	0	0	1	1
1	0	1	0	1	1
1	0	0	0	0	1
1	1	1	1	1	1

The design shows the lower case letter d.

PRACTICE QUESTION

2 A font designer creates some simple letter designs as bitstreams. In each case the pixel grid was 8 × 8.

Give the letters shown in the bitmaps represented by the following bitstreams.

a 11111111 / 11000111 / 10010011 / 10111011 / 10000011 / 10111111 / 10010011 / 11000111

b 11111111 / 10000011 / 10111111 / 10111111 / 10000111 / 10111111 / 10111111 / 10000011

c 11111111 / 11000111 / 10010011 / 10111011 / 10111011 / 10111011 / 10010011 / 11000111

d 11111111 / 10000111 / 11011011 / 11011011 / 11000111 / 11011011 / 11011011 / 10000111

Extra bits

Sometimes computer systems use additional bits to record the bitmap. For example, there might be extra bits 'padding out' the start and end of each line. In cases where this is important, the information you need will be provided.

STRETCH YOURSELF!

Design some letters in a font of your own design and represent them as a bitstream.

REMEMBER: To change a bitstream into an image:

1 find the size of the grid and draw it

2 insert the series of 1s and 0s into the grid

3 shade cells containing 0.

4.3 Representing colour

Colour codes

All data must be held in a computer as electronic binary numbers.

A computer can store coloured images using number codes which stand for different colours. There are several coding systems used to store information about colour and different image file formats use different colour coding systems.

Colour codes and pixels

The computer stores information about each pixel in a bitmap image. In Topic 4.1 you learnt that in a monochrome image each pixel is represented by a single bit. In a multi-coloured image there are more than two colour choices, so the information must be stored as a longer code number.

Often, the colour of each pixel is stored by one code number. A whole coloured image is stored in a computer as a series of code numbers.

8-bit colour

8-bit colour is a system where eight bits – or one byte – of data is used to store one pixel. From your study of binary numbers you will know that one byte can store all the numbers from 0 to 255.

This means that an image made using 8-bit colour can have up to 256 different colours. In real life there are far more than 256 different colours. That means an 8-bit image may not have clear and realistic colours.

Using fewer bits to store colour will give an even smaller range of colours.

WORKED EXAMPLE

Calculate how many colours are available in an image format that uses four bits.

The largest number that can be represented by four bits is 1111. Putting this in the binary position grid:

8	4	2	1
1	1	1	1

$$8 + 4 + 2 + 1 = 15$$

With four bits the computer can store all the numbers up to 15, including the number 0. That is 16 values altogether so there are 16 colours available.

PRACTICE QUESTION

1 State how many colours are available in these file formats.

- **a** 2-bit colour
- **b** 4-bit colour
- **c** 5-bit colour
- **d** 6-bit colour

File size

In an 8-bit colour image, one pixel takes one byte of storage. To find the size of an image file in 8-bit colour you need to find how many pixels there are in the image. This is the number of bytes needed to store the image.

You may need to simplify the expression using suitable digital units (see Topic 3.2).

PRACTICE QUESTION

2 Calculate how much storage space is taken up by the following images, stored using 8-bit colour.

a an image with 30 000 pixels

b an image with 3 800 000 pixels

c an image with a pixel grid of 640 × 480

d an image with a pixel grid of 1200 × 1800

Use suitable units to express your answer.

Other file formats

To find the file size of an image in bits:

- calculate the number of pixels in the image
- multiply by the number of bits in the colour code.

To turn this result into bytes:

- divide the number of bits by eight to give the total number of bytes.

Express this using the most suitable digital units.

WORKED EXAMPLE

Calculate how many bytes it takes to store a 4-bit image of size 320 × 240 pixels.

The total number of pixels is the image width × height.

$320 \times 240 = 76800$

The image has 76 800 pixels. Each pixel is stored using 4 bits.

$76800 \times 4 = 307200$

The image is stored using 307 200 bits. To convert into bytes divide by eight.

$$\frac{307\,200}{8} = 38\,400$$

So the image uses 38 400 bytes. We can express this in its simplest form as:

38.4 kB

PRACTICE QUESTION

3 Calculate how much storage space is taken up by the following colour images.

a an image with a pixel grid of 640 × 480 in 6-bit colour

b an image with a pixel grid of 1200 × 1800 in 2-bit colour

c an image with a pixel grid of 640 × 480 in 12-bit colour

d an image with a pixel grid of 1200 × 1800 in 16-bit colour

REMEMBER: To find the size of an image file, multiply the number of bits in the colour code by the number of pixels. This gives you an answer in bits.

4.4 Colour depth

Colour depth

The number of colours available to make an image is called the colour depth.
Image formats with high colour depth:

- have long colour code numbers
- can use a wide range of different colours
- take up more storage space.

Images with low colour depth:

- have short colour code numbers
- have a smaller range of colours
- take up less storage space.

Quality trade-off

The size of an image file depends on:

- how many pixels there are in the image
- how big the colour codes are.

There is a trade-off between the colour depth of an image and the file size.

Some image formats use more than 1 byte to store the colour code. This increases the number of available colour codes but makes the file size larger.

WORKED EXAMPLE

Calculate how many colours are available in 16-bit colour.

There are eight bits in a byte. So 16 bits is equivalent to two bytes. Each byte can store 256 different codes. So two bytes can store:

$256 \times 256 = 65536$

So, using 16-bit colour in an image means the image can use more than 65 thousand different colours, giving a much greater depth of colour.

PRACTICE QUESTION

1 Complete the table to show many colours are available in these file formats.

Colour format	Number of bytes for each pixel	How many colours?
8 bit	1	
16 bit		$256 \times 256 = 65536$
24 bit	3	
32 bit		

True colour

24-bit colour is now used by almost every computer and phone display and most image storage formats. 24-bit colour can store over 16 million different colours. This is about the same as the number of different colours that the human eye can see. It gives images a very realistic colour depth so 24-bit colour is sometimes called true colour.

RGB colour

Images in 24-bit colour use three bytes to store each colour code. They typically do this using the RGB system. RGB stands for red, green, blue. The three bytes of the colour code are used to store three different values.

- The first byte stores the amount of red light in the pixel.
- The second byte stores the amount of green light in the pixel.
- The third byte stores the amount of blue light in the pixel.

Each colour can have a brightness set from 0 to 255. 0 is the darkest (no light of that colour). 255 is the brightest (maximum light of that colour). By mixing the three types of light in different intensities, a huge range of different colours can be made.

Table 1 Examples of RGB colour codes

Colour code	What colour does it show?
0, 0, 0	black
255, 0, 0	red
128, 0, 128	purple
255, 255, 0	yellow
255, 255, 255	white

When writing RGB values, using hexadecimal can be useful for two reasons.

- It is easy to convert between hexadecimal and binary.
- The number is shorter than if we used decimal.

32-bit colour

32-bit colour has a fourth byte as well as the three bytes used for RGB values. This extra byte is sometimes used to store the transparency of the colour. It is used when images are laid on top of each other in a composition. Sometimes the extra byte is left unused.

Calculate file size

To calculate file size:

- multiply the number of bits in the code by the number of pixels
- divide by eight to give the number of bytes in the file.

PRACTICE QUESTION

2 Complete the table to show the size of these images.

Image size	Colour depth	Number of bytes
1200 × 1800	24-bit colour	
240 × 360	16-bit colour	
640 × 480	32-bit colour	
2400 × 3200	true colour	

REMEMBER: Colour depth tells you the number of bits used to store each pixel in an image. The greater the number of bits, the wider the range of colours that can be used in the image.

4.5 Image resolution

How many pixels?

An image can be stored using lots of small pixels or a few larger pixels. The number of pixels in the image is called the resolution.

- An image made with lots of small pixels is called high resolution.
- An image made with fewer, larger, pixels is called low resolution.

A high resolution image has more detail. The image is clearer and more realistic.

Figure 1 Comparison between a low-resolution (left) and a high-resolution image (right)

A low resolution image has less information. It has less detail and is often less realistic.

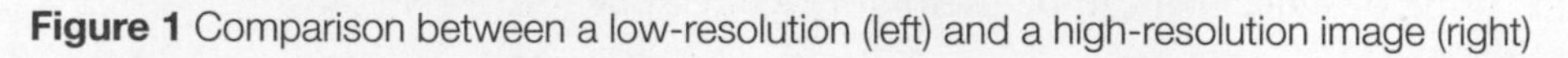

PRACTICE QUESTION

1 Compare the two images in Figure 1.

a State which of the images is low resolution.

b It is easy to see the advantage of a high resolution image. Give the advantages of a low resolution image.

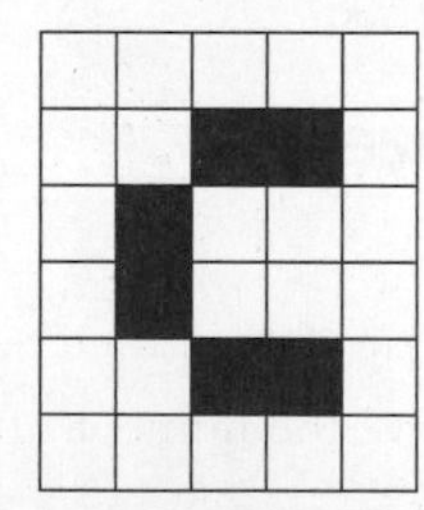

The resolution is fixed by the number of pixel codes in the image file. If the image is displayed on a larger screen the number of pixels in the image will not increase; the resolution stays the same. That is because the number of colour codes in the image file will not change. There is no new information coming into the computer.

For example, this simple image represents the letter 'c'. It is stored in a grid of width 5 × height 6 = 30 pixels. Each pixel provides the colour value for one point of light on the computer screen.

In this example the image has been expanded to twice the size. Each pixel provides the value for four points of light on the computer screen. The amount of detail has not changed.

No extra information has been added to the file so making the image larger does not make it any clearer: it does not increase the resolution of the image.

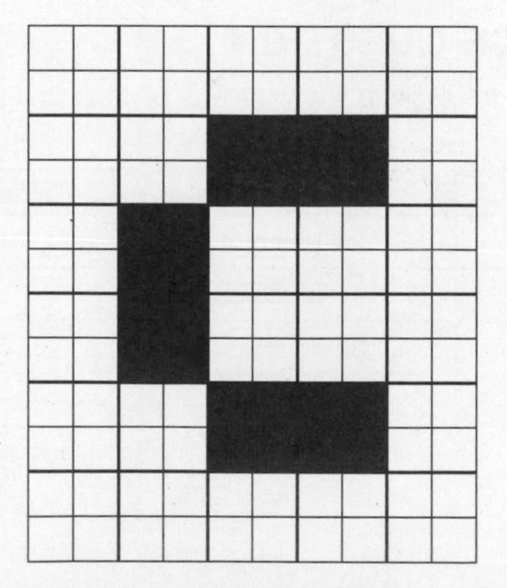

PRACTICE QUESTION

2 a Calculate how many bits of computer storage would be needed to store these images of the letter 'c' as monochrome bitstreams.

b Both of these images use the same amount of storage. Calculate how many bytes of storage are used. (Round up to the nearest byte.)

c Calculate how many bytes of storage would be needed if these images were stored using 32-bit colour.

What is a pixel?

A pixel is not a physical property of the computer screen. It is the smallest editable part of an image. It is a piece of information that lets the computer set the value of one point in the image.

When an image is made larger, the pixels get bigger. But no extra pixels, or details, are added to the image.

Summary: calculate file size

To calculate the file size of a bitmap image, multiply three values:

- the width of the image in pixels
- the height of the image in pixels
- the depth of colour in bits.

This gives the file size in bits. To show the size in bytes:

- divide by eight to give the file size in bytes
- choose a suitable digital unit if the number of bytes is large.

Metadata

The size of a file may be larger than the result of this calculation because images files can contain additional information called metadata. Metadata means data about the image. It is usually text information.

For example, if you take a photo with a digital camera or smartphone the photo file might include:

- information added automatically by the processor inside the camera or phone, such as the date and time when the photo was taken and the location of the shot
- information entered by the photographer, such as the names of the people in the image
- legal information, such as copyright and the email address of the photographer.

Metadata usually takes up very little space compared to the size of the photo image.

PRACTICE QUESTION

3 Calculate the file size of the following images.

a 240 × 480 pixels, using 32-bit colour with 30 kB of metadata.

b 1200 × 3000 pixels, using 16-bit colour with no metadata.

c 36 × 36 pixels using 8-bit colour and 4 kB of metadata.

d 1800 × 1800 pixels, using 24-bit colour with no metadata.

REMEMBER: To find the size of a file, multiply image width and height to find the number of pixels. Multiply by colour depth to give the size in bits. Divide this by eight to give the size in bytes. Finally, add the size of any metadata.

5 DIGITAL SOUND

5.1 The nature of sound

Sound waves

Sound is how the human ear interprets waves of sound pressure that move through the air around us. Sound waves are typically drawn as a curved line.

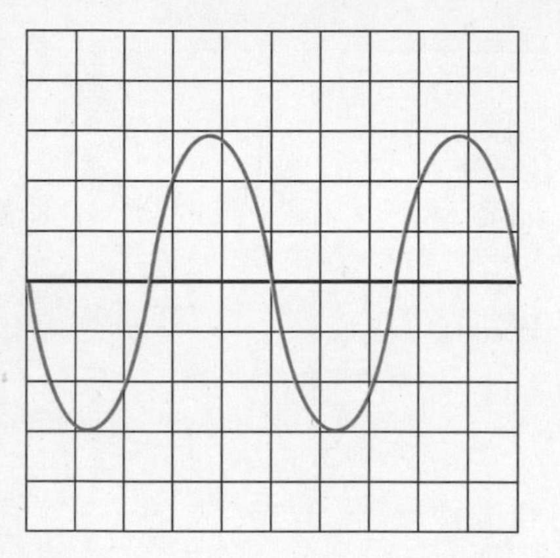

Figure 1 An analogue signal varies continuously

The diagram above is a graph of changes in air pressure over time. The changes make a smooth curve. They are called waves, by analogy with waves on the surface of water.

Analogue measurements

Air pressure is an example of an analogue measurement. An analogue measurement is one that varies or changes smoothly between different levels. Air pressure can change quickly, but it does not jump straight from high to low. Instead it passes through all the levels of pressure in between the two states.

An analogue measurement can never be completely expressed as a number, even using very precise fractions.

PRACTICE QUESTION

1 a State how many sound waves are shown in Figure 1.

b Give the measurement represented by the X-axis of the graph.

c Give the measurement represented by the Y-axis of the graph.

Frequency and amplitude

Sound waves have two important characteristics, frequency and amplitude.

- *Frequency* measures how many sound waves reach your ear in 1 second. As sound travels at a constant speed, this depends on how close together the sound waves are (the wavelength).
- *Amplitude* measures the maximum height of each sound wave on a graph. This is a measure of the change in air pressure as it increases and decreases. The distance of the line from the baseline in the centre of the graph represents the amplitude.

The human ear can detect the frequency and amplitude of sound waves.

- The *pitch* of a note is based on its frequency: high frequency is heard as a high note, low frequency as a low note.
- The *loudness* of a sound is based on its amplitude: the bigger the amplitude, the louder the sound.

Measuring frequency and amplitude

Frequency is measured in hertz. 1 hertz means repeated once a second, 2 hertz means repeated twice a second, and so on. The lowest frequency sound a person can hear is about 20 hertz. The highest frequency sound a person can hear is about 20 000 hertz (also written as 20 kilohertz).

Amplitude is measured in decibels. A whisper is about 15 decibels and loud fireworks are about 150 decibels.

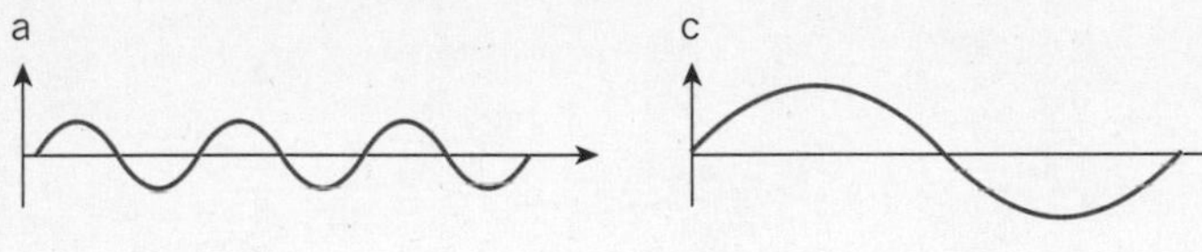

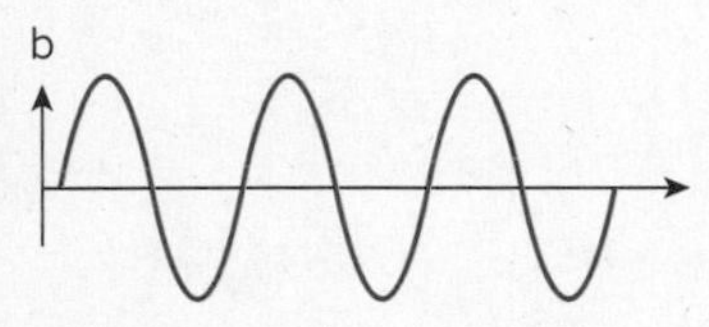

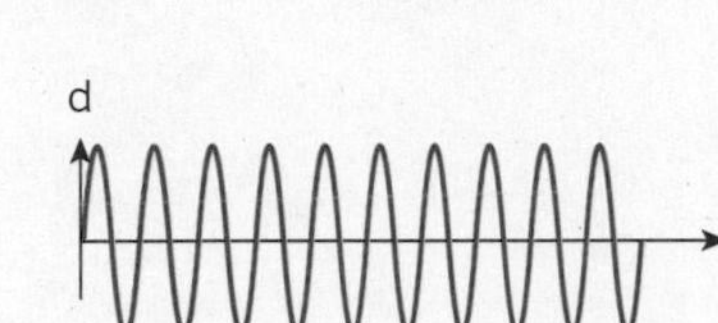

Figure 2 Graphs showing sound waves with various amplitudes and frequencies

PRACTICE QUESTIONS

2 Figure 2 shows four graphs of sound waves.

- **a** State which image out of a and b shows the higher amplitude.
- **b** State which image out of c and d shows the higher frequency.
- **c** State which of the four sounds would be the quietest.
- **d** State which of the four sounds would be the highest in pitch.

3 The waves in Figure 2 show the changes in air pressure during $\frac{1}{10}$ of a second. Give the frequency, in hertz, of:

- **a** sound a
- **b** sound b
- **c** sound c
- **d** sound d
- **e** Suggest which sound would be impossible for a human to hear.

REMEMBER: Sound waves have frequency and amplitude. Frequency and amplitude are analogue measurements. An analogue measurement is one that varies or changes smoothly between different levels.

5.2 Digital recording

Analogue-to-digital conversion

All data has to be turned into number form when it is input to a computer. That includes analogue measurements such as air pressure – the basis of sound.

Recording analogue data on a digital device involves analogue-to-digital conversion. This means turning an analogue measurement, which changes smoothly and continuously, into a series of separate number values.
Figure 1 shows this conversion visually.

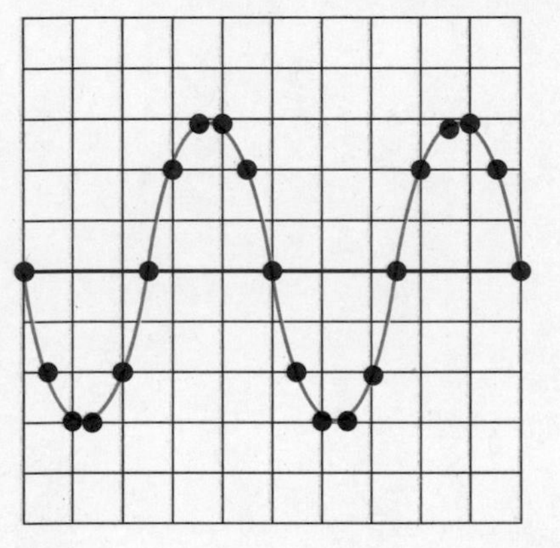

Figure 1 An analogue signal is sampled at intervals

Each dot on the graph represents a sample taken by the computer. The sound wave is sampled at regular intervals. Each sample produces a single number value that represents the amplitude (the distance between each point and the baseline).

The amplitude values are stored as a series of binary numbers in the computer system.

Sampling

A continuous analogue signal is converted into a series of separate digital values. This involves sampling the signal over time. The series of sample numbers is not the same as the analogue measurement. But if the samples are close enough together it produces a good copy of the signal.

The dots on the graph give a good indication of the shape of the sound wave. Similarly, when many samples are taken over time, the series of numbers will record the sound.

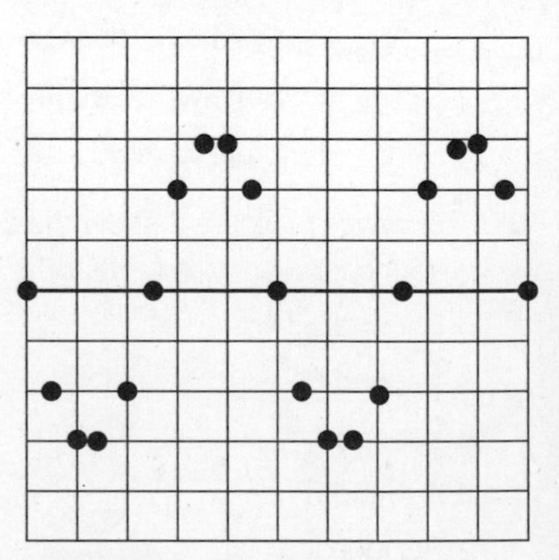

Figure 2 The digital sample is an approximation of the analogue signal

Sample rate

The sample rate is a measure of how many samples are taken in 1 second. The more samples there are, the better the quality of the sound.

Sample rate is measured in hertz. You have already learnt that hertz is a measurement of how often something is repeated.

- If a sample were taken once a second that would be a sampling rate of 1 hertz (1 Hz).
- If the sample were taken ten thousand times a second that would be 10 kilohertz (10 kHz).

PRACTICE QUESTION

1 **a** State how many sample points are shown in Figure 1.

b If this graph represents $\frac{1}{1000}$ of a second, give the number of samples taken in 1 second.

c Give the sampling rate in hertz. Use a suitable unit to express your answer.

Sample rate and sound quality

The more samples there are in a digital recording of a sound, the more closely it will match the analogue sound. With fewer samples, the sound will be distorted and hard to understand. If the sampling rate is too infrequent then the recording will be of a poorer quality.

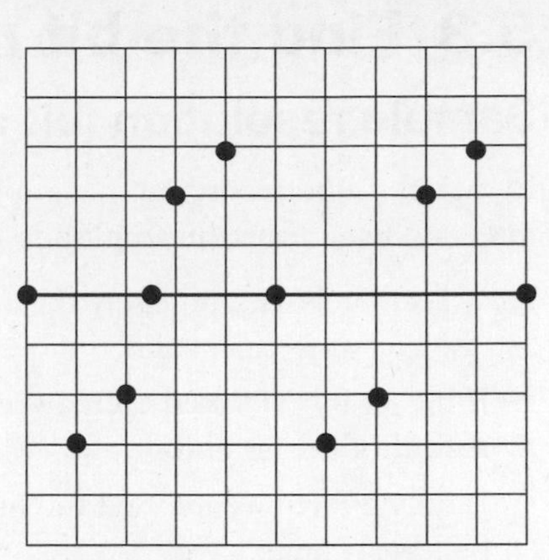

Figure 3 A low sample rate – resulting in a poor quality recording

WORKED EXAMPLE

A telephone line transmits sound signals at a rate of 8 kHz. Calculate the number of sound samples transmitted in $\frac{1}{100}$ of a second.

1 hertz = once per second. 1 kHz = a thousand times per second.

A rate of 8 kHz = the sound is sampled 8000 times per second.

To find the number of samples in $\frac{1}{100}$ of a second, divide by 100

$$\frac{8000}{100} = 80$$

So a phone line signal samples the sound 80 times in $\frac{1}{100}$ of a second.

PRACTICE QUESTION

2 **a** In a music studio the recording devices sample the sound 44 100 times per second. Express this sample rate using a suitable unit.

b A telephone line sends sound signals at a rate of 8 kHz. Explain what this tells you about the sound quality of a telephone compared to the sound quality of a recording studio.

c A sampling rate of 11.025 kHz provides a clear recording of human speech. Calculate how many samples there would be in a 10-minute recording of an audio book at this sampling rate.

d Explain why the normal sampling rate for recording music might be higher than the normal sampling rate for understanding speech.

REMEMBER: Analogue measurements must be sampled to turn them into digital measurements.

The amplitude of the sound is measured at regular intervals and converted into a digital value.

Sampling rate is measured in hertz. The number of hertz is the number of samples per second.

5.3 Find the bit rate

Sample resolution (bit depth)

In digital sound recording the amplitude of the sound wave is sampled at regular intervals. Each time the amplitude is sampled this value is recorded as a number.

How accurately and precisely the amplitude is recorded depends on how many bits are used to store each value.

- If lots of bits are used then a very precise number can be stored, giving an accurate record of the amplitude.
- If only a few bits are used the number will have to be rounded up or down, it may be less accurate.

The number of bits used to record sound is known as the bit depth or the sample resolution.

Greater sample resolution gives a better sound quality. Of course using a lot of bits to store each sample makes the file size larger, so there is a trade-off between the sound quality and the size of the sound file.

Sound formats

Different sound formats use different sample resolution. Some examples are shown in the next table.

Sound format	Sample resolution	How many code numbers
NICAM (used for TV broadcast)	10 bits	1024
CD	16 bits	126 × 126 =
Blu-Ray	24 bits	

PRACTICE QUESTION

1 **a** Complete the final column of the table to show how many code numbers are available to record amplitude in each sound format.

b Explain which format you would expect to provide the most accurate and realistic sound quality.

Bit rate

The bit rate is the number of bits it takes to store one second of sound.

Calculate the bit rate by multiplying:

- the number of samples taken in one second (the sampling rate in hertz)
- the number of bits used to store one sample (the sample resolution).

The result is a number of bits.

WORKED EXAMPLE

Calculate the bit rate for a music recording with a sample rate of 44.1 kHz and a sample resolution of 16 bits.

The sampling rate is 44.1 kHz. Convert this into Hz:

44.1 kHz = 44 100 Hz

Multiply this by the sample resolution:

44 100 × 16 = 705 600

The result is 705 600 bits to store 1 second of music.

PRACTICE QUESTION

2 Calculate the bit rate for the following sound samples.

a sample rate 20 kHz, sample resolution 16 bits

b sample rate 8 kHz, sample resolution 12 bits

c sample rate 44.1 kHz, sample resolution 24 bits

d sample rate 11.5 kHz, sample resolution 8 bits

Stereophonic sound and more

A stereophonic (stereo) track is made of two recordings. The recordings are taken from different positions at the same time. When the recording is played, the two tracks play from different speakers. This gives a more realistic sound output. A stereo recording has two tracks so the bit rate must be multiplied by two.

Another version, quadrophonic sound, uses four recording devices. The four tracks are played from four different speakers placed all around the room. This is used in many cinemas and is also called surround sound. If a track is recorded in quadrophonic surround sound the bit rate must be multiplied by four.

WORKED EXAMPLE

A DVD movie includes a quadrophonic surround sound soundtrack with a sample rate of 44.1 kHz and a bit depth of 24. Calculate the bit rate.

The sampling rate is 44.1 kHz. Convert this into Hz:

44.1 kHz = 44 100 Hz

Multiply this by the sample resolution:

44 100 × 24 = 1 058 400

Then multiply by four because quadrophonic sound has four audio tracks:

1 058 400 × 4 = 4 233 600

So the DVD uses 4 233 600 bits per second of soundtrack.

PRACTICE QUESTION

3 Calculate the bit rate for the following sound samples.

a sample rate 20 kHz, sample resolution 16, stereophonic sound (two tracks)

b sample rate 11 kHz, sample resolution 10, stereophonic

c sample rate 44.1 kHz, sample resolution 32, quadrophonic (four tracks)

d sample rate 8 kHz, sample resolution 12, quadrophonic

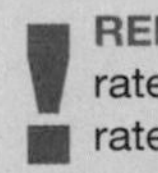

REMEMBER: To find bit rate multiply sampling rate by sample resolution.

The bit rate is a value in bits not in bytes.

5.4 Audio file size

Length of recording

You have learnt how to calculate the bit rate of an audio track. This depends on the sampling rate and the sample resolution (also called bit depth). The bit rate tells you the number of bits used to record 1 second of sound. If the sound is stereophonic this must be doubled.

To work out the total file size of a recording you will need to know how many seconds the recording lasts. If you are given the length of the recording in minutes, remember to multiply by 60 to convert to seconds.

total number of bits = bit rate per second × number of seconds

The result is a value expressed as bits.

WORKED EXAMPLE

A music recording has a sampling rate of 44.1 kHz and a sample resolution of 16 bits. The recording uses stereophonic sound and lasts three minutes. Calculate how many bits are in the recording.

Multiply sampling rate by sample resolution:

$44\,100 \times 16 = 705\,600$

Multiply by two as the recording is in stereo:

$705\,600 \times 2 = 1\,411\,200$

This gives the bit rate of 1 411 200.

The recording is 3 minutes long. In seconds this is:

$3 \times 60 = 360$

Multiply the bit rate by the number of seconds in the recording:

$1\,411\,200 \times 360 = 508\,032\,000$

The total number of bits in the recording is 508 032 000 (more than half a billion bits).

PRACTICE QUESTION

1 Calculate the number of bits in the following recordings.

- a a 10-second sound recording at a bit rate of 200 000
- b a 10-minute sound recording at a bit rate of 88 000
- c a 45-second recording with a sample rate of 44.1 kHz and a bit depth of 16
- d a 2-minute stereophonic recording with a sample rate of 32 kHz and a bit depth of 12

Convert into bytes

The previous calculation gives you a result in bits. File size is generally expressed in bytes. In most cases it is best to divide the number of bits by eight to give a file size in bytes.

Audio files are generally quite large. Use a suitable digital unit to represent the size in a way that is easy to read and understand.

WORKED EXAMPLE

The number of bits in an audio recording is 508 032 000. Express this value as a file size using appropriate digital units.

The recording contains 508 032 000 bits. Divide by eight to give the number of bytes.

$$\frac{508\,032\,000}{8} = 63\,504\,000 \text{ bytes}$$

The file will contain 63 504 000 bytes.

This value is best expressed in megabytes. 1 megabyte is 1 million bytes, so:

63 504 000 bytes = 63.504 MB

In practice this might be rounded down to 63.5 MB or up to 64 MB.

PRACTICE QUESTION

2 Calculate the file size of the following sound recordings.

- **a** a 2-minute sound recording at a bit rate of 360 000
- **b** a 1-hour sound recording at a bit rate of 88 000
- **c** a 20-second recording with a sample rate of 44.1 kHz and a sample resolution of 24 bits
- **d** a 7-minute recording with a sample rate of 11 kHz and a sample resolution of 16 bits

Remember file size is shown in bytes not bits. Use a suitable unit to express the value.

The quality/file size trade-off

The quality of an audio recording can be improved by:

- increasing the sampling rate so that the digital series of numbers is closer to the shape of the analogue signal
- using a larger bit depth so the amplitude of the sound can be recorded with greater precision and improved accuracy.

Because bit rate is calculated by multiplying sampling rate by bit depth, increasing either of these will increase bit rate. This will increase file size, so in audio recording there is a trade-off between file size and file quality.

REMEMBER: To find audio file size, multiply the bit rate by the number of seconds. This gives the size in bits. Divide by eight to give the size in bytes.

6.1 The importance of file size

File size is measured in bytes or other digital units, such as megabytes. Many factors affect the size of a data file. In general smaller data files have many advantages, for example, taking up less storage space and being faster to transmit via network connections.

Small files and storage

Is taking up less storage space important? The cost of digital storage space is not expensive nowadays. Ordinary computer users can buy storage devices, such as flash drives and hard disks, that can store lots of files.

Many people want to save lots of files, such as family photos and videos of important events. People prefer small file sizes so that they do not run out of space.

Another storage solution that people use is cloud storage. Cloud storage is storage you can access through an Internet connection. Cloud storage is often free up to a certain limit. Users want to stay inside this limit, meaning small file sizes are preferred.

WORKED EXAMPLE

A cloud storage service offers up to 10 GB of free storage space. Calculate how many files you can store if the average file size is 400 MB.

Convert the storage and file sizes to the same units. Megabytes are most suitable.

10 GB = 10 000 MB

Divide the amount of available storage by the average file size.

$$\frac{10\,000}{400} = 25$$

So 25 files can be stored for free.

PRACTICE QUESTION

1 A music student wants to send his tutor examples of electronic music on a CD. He is worried about whether he has enough storage space.

Calculate the file sizes of the following tracks.

- **a** a 300-second track at a bit rate of 128 000 bits per second
- **b** a 2.4-minute track at bit rate of 160 000 bits per second
- **c** a 10-minute track at a bit rate of 256 000 bits per second
- **d** Calculate how much storage space he needs in total.

Remember to convert file sizes to bytes or other suitable units.

Download speed

Internet users also prefer small file sizes. 'Downloading' a file means copying it to your computer over an Internet connection. If the file is small it will download quickly. If the file is large it may take a long time to download.

Quality and file size

You have learnt how image and sound data is held in the computer. Data can be held at higher or lower quality levels.

In general, higher file quality means recording more data, storing more values, and storing larger values. In higher quality files:

- the sampling or resolution is higher
- larger code numbers are used to represent colours or sounds.

This means that there is always a trade-off between file size and file quality. In general, higher quality files are larger. They take up more storage space.

PRACTICE QUESTION

2 A photo of a comet was available in these different image formats. In each case estimate the file size used to store the photo. Give your answers using suitable units.

a 16-bit colour; 4000 pixels

b monochrome colour; 2000 pixels

c 24-bit colour; 160 × 100 pixels

d Rank the photos a, b, and c in order of probable image quality. Explain your answer.

REMEMBER: Small files save storage space. Small files are faster to download. But high quality image and sound files are larger in size.

STRETCH YOURSELF!

Larger files take longer to download. If a web site contains large files, for example images, it will take longer to load. It will also take longer on a slow Internet connection.

The speed of an Internet connection is measured in bits per second. Notice that it is bits rather than bytes.

Calculate how long would it take to download an image file of 14.4 MB at a download speed of 768 kilobits per second?

First calculate the number of bytes per second

768 kilobits per second means 768 000 bits can be downloaded in one second.

Divide that by eight to give the number of bytes per second.

$$\frac{768\,000}{8} = 96\,000 \text{ bytes}$$

So at this speed you can download 96 kB per second.

Then divide the size of the file – 14.4 MB – by 96 kB to get the number of seconds.

First put the two values into the same units. 14.4 MB is 14 400 kB

$$\frac{14{,}400 \text{ kB}}{96 \text{ kB}} = 150$$

So it would take 150 seconds, or 2 minutes 30 seconds

6.2 Lossless and lossy compression

Types of compression

Compression is a way of making large files smaller. The two main types of compression are lossless and lossy compression.

- *Lossy* compression means making files smaller with some loss of data quality.
- *Lossless* compression means making files smaller with no loss of data quality.

In general, lossy compression has a bigger effect on file size, but it has the disadvantage of making the data quality worse.

Lossy compression

Lossy compression works by reducing data quality, for example, by:

- reducing the resolution and colour depth of an image
- reducing the sampling rate and bit depth of an audio recording.

The effect of these changes will be to lose information. The sounds and images will often be less clear and detailed.

WORKED EXAMPLE

An image was created using 32-bit colour. The image size was 300 × 240 pixels. The image was compressed by changing it to 8-bit colour. Calculate the change in file size.

First work out the file size of the original image. The image is 300 pixels wide and 240 pixels tall.

$$300 \times 240 = 72\,000 \text{ pixels}$$

Each pixel is stored using 32 bits.

$$72\,000 \times 32 = 2\,304\,000 \text{ bits}$$

So the image uses 2 304 000 bits. Turn this into bytes.

$$\frac{2\,304\,000}{8} = 288\,000 \text{ bytes}$$

This value is best expressed as 288 kB.

The second version of the file has the same number of pixels, but each pixel is stored using eight bits (i.e., one byte). This means there are exactly the same numbers of bytes as pixels.

$$72\,000 \text{ pixels} = 72\,000 \text{ bytes} = 72 \text{ kB}$$

The two file sizes are:

- original file = 288 kB
- compressed file = 72 kB.

The difference in file size can be worked out by subtracting one from the other.

$$288 - 72 = 216 \text{ kB}$$

This form of compression has reduced the file size by 216 kB; It has gone down from 288 kB to 72 kB. What has happened to the file quality?

PRACTICE QUESTION

1 **a** A 24-bit colour image has 2000 × 2400 pixels. The resolution is reduced to 1200 × 1440 pixels. Calculate the reduction in file size.

b The file is further compressed by making it monochrome. Calculate the new file size.

c A 10-minute audio recording has a sample rate of 44.1 kHz, bit depth (sample resolution) of 24 bits, and is recorded in stereo. In order to compress the file, the bit depth is reduced to 16 and the stereo effect is removed. Calculate the reduction in file size.

d The audio file is further compressed by reducing the sample rate to 11 kHz. Calculate the new file size.

WORKED EXAMPLE

An image has 40 000 pixels and uses 24-bit colour. Calculate the file size. Calculate the percentage reduction of the file size of reducing this to 16-bit colour.

Each pixel uses 24 bits. $\frac{24}{8} = 3$. So each pixel uses 3 bytes.

$40000 \times 3 = 120000$

The original image file is 120 kB.

Calculate the percentage reduction of the file size of reducing this image to 16-bit colour.

16-bit colour uses 2 bytes per pixel.

$40000 \times 2 = 80000$

The compressed image file is 80 kB so the reduction is 40 kB.

$$\text{percentage reduction} = \frac{40}{120} = \frac{1}{3} = 33\%$$

PRACTICE QUESTION

2 **a** An image was stored using 520 × 640 pixels. The colour depth was 16-bit. The resolution was reduced to 260 × 320. The colour depth was reduced to 8-bit colour. Calculate the reduction in file size.

b A 10-second audio file had a sample rate of 44.1 kHz. The sample resolution was 16-bit. The sample rate was reduced to 24 kHz, but the sample resolution was not changed. Calculate the reduction in file size.

c Calculate the percentage reduction in file size if an audio file is changed from 16-bit to 12-bit sample resolution and all other factors remain the same.

d Calculate the percentage reduction in file size if an image is changed from 24-bit colour to monochrome and all other factors remain the same. Give the percentage reduction to the nearest whole number.

REMEMBER:
Compression reduces the size of a file saving storage space and making downloads faster. Lossy compression works by reducing data quality. There are fewer data points – or there is a smaller range of colour or sound.

6.3 Run-length encoding

Lossless compression

You have learnt about methods that reduce file size by reducing data quality. This is called **lossy** compression.

It is sometimes possible to reduce file size without loss of data quality. This is called **lossless** compression. Typically lossless compression does not produce such a large reduction in file size as lossy compression.

Lossless compression takes advantage of repetition of values in a data file. You will learn about two important methods of lossless compression:

- run-length encoding
- dictionary coding.

These methods of compression may be used with different types of data, including image and audio data.

Run-length encoding (RLE)

Many data files include large blocks of repetitive data. For example, an image file may have a large group of pixels that all have the same colour value. Think of the logos of your favourite products. They probably use limited colours and no colour shading.

In a normal bitmap file each pixel will be stored as a colour code. If there are 1000 pixels of the same colour, then the code number will be repeated 1000 times.

RLE removes this repetition to save storage space. In RLE:

- the colour code is stored once
- a number is added that tells the computer how many times to repeat the colour.

RLE works where there are large blocks of repeated data.

WORKED EXAMPLE

Here is a block of text data. Show how RLE can be used to reduce the data size.

aaaabbbbbbbbbbbdddddddddwwwwwwwwwwwwkkkkkkkkkkkdddddpppppppppp

Split the data to show the repeated values:

aaaa

bbbbbbbbbbb

ddddddddd

wwwwwwwwwwww

kkkkkkkkkkk

ddddd

pppppppppp

Next, count how many times the data is repeated.

Original data	Data value	Frequency
aaaa	a	4
bbbbbbbbbbb	b	11
ddddddddd	d	9
wwwwwwwwwwww	w	12
kkkkkkkkkkk	k	11
ddddd	d	5
pppppppppp	p	10

Now the data can be represented using the data/frequency pairs:

a-4 b-11 d-9 w-12 k-11 d-5 p-10

There are different ways of writing the pairs. As long as it is clear what the data value and the frequency value are, you have expressed the data correctly.

PRACTICE QUESTION

1 Represent the following data sets as data/frequency pairs using RLE.

a *********&&&&&&&&&&&&$$$$$$$$$$$$$$$%%%%%%%%%?????

b ..\\\......\............@@@@@@@@@@@@@@\\\\\\\\\\\\\

c SSSSSSSSWWWWWWWWWWWWWWWEEEEEEEEEETTTTTTTTTTTTTT

d ###############33333333333##############99999999999999999

WORKED EXAMPLE

A solid red background on a website was stored using 24-bit colour (RGB values). The image is made of a series of 1700 pixels that all have the colour code 200, 0, 50.

Calculate how much storage space could be saved if the file is stored using RLE.

First, find the original file size. Storing each colour code takes 24 bits, or three bytes, of data. There are 1700 pixels. So the total amount of storage would be:

$1700 \times 3 = 5100$ bytes (5.1 kB)

Then find the storage space using RLE. The colour code is stored. Then the frequency is recorded.

The colour code is 200, 0, 50. This takes 3 bytes.

The frequency is 1700. Storing this number takes 2 bytes.

$2 + 3 = 5$ bytes

Finally, calculate the difference. The file size has been reduced from 5100 bytes to 5 bytes with no loss of data quality.

$5100 - 5 = 5095$ bytes

PRACTICE QUESTION

2 Calculate how much of a reduction in file size can be achieved by compressing the following strings using RLE.

a *********&&&&&&&&&&&&$$$$$$$$$$$$$$$%%%%%%%%
%%??????????

b ..\\\......\............@@@@@@@@@@@@@@\\\\\\\\\\\\\

c SSSSSSSSWWWWWWWWWWWWWWWEEEEEEEEEE
TTTTTTTTTTTTTT

d ###############33333333333##############99999999
999999999

> **REMEMBER:** Run-length encoding is also called RLE. It is used on files that have large blocks of repeated data. Instead of repeating each data value, the value is stored once, together with a number representing the frequency of the value.

When to use RLE

- RLE can be used with any data file that includes sequences where the same value is repeated in a continuous block.
- RLE cannot be used if a data file does not include sequences of the same value. If every pixel in an image is a different colour from the one next to it, then RLE will not compress the file.

6.4 Dictionary compression

Limitations of RLE

Run-length encoding produces good compression results if a file has large blocks of repeated data. However, many files do not have this feature. A method called dictionary compression can be used instead.

NOTE: File compression software might combine *both* dictionary compression and RLE. Using both methods gives a good result and works on a wide range of files.

Dictionary compression – an alternative to RLE

Like RLE, dictionary compression reduces file size with no loss of file quality. Dictionary compression works if a file has repeated values that are scattered about in the file. The values do not need to be in big blocks.

For example, in a realistic photo it is rare to see big blocks of the same colour, but it is quite common to see the same colour repeated in different places.

Dictionary compression:

- finds the most common values used in the file
- makes a dictionary by assigning a short code to each common data value
- goes through the file substituting the short dictionary code for the corresponding data value.

The short code takes up less file space than the original value so the file becomes much smaller.

WORKED EXAMPLE

DNA is represented using the letters ACGT only. A file contained the following text data. It was stored using ASCII code.

TGGTCCCCTAAGTTACGTAGGGTGCCGGGCAC

Explain how you would use dictionary coding to compress the file.

There are only four values stored in the text string – A, C, G, and T. In the original file each letter is stored using ASCII code. Here are the binary ASCII codes for the letters.

Value	ASCII code
A	0100 0001
C	0100 0011
G	0100 0111
T	0101 0100

ASCII code uses 1 byte (or 8 bits) to store each character.

We can create a simple dictionary like this. Each value is given a short binary code number.

Value	ASCII code	Dictionary code
A	0100 0001	00
C	0100 0011	01
G	0100 0111	10
T	0101 0100	11

The new code uses 2 bits to store each character. Using this shorter code will use less space without losing any of the information in the file. There is therefore no loss of file quality.

PRACTICE QUESTION

1 A technology college has 2000 students who are each enrolled on one of eight available courses. The name of each course is 20 characters long. A file is created that stores each student's name and the name of the student's course.

a Give the number of bytes used to store each course name.

b If there are 2000 course names in the database, calculate the total amount of storage used to record course names.

c A dictionary was created to store course names using a short code. Calculate how many bits are needed to give each course a unique code.

d Calculate the total amount of storage used to record 2000 course codes.

e Calculate the reduction in file size achieved by using dictionary encoding.

NOTE: Using a dictionary can produce a big reduction in file size. Dictionary compression works best when the file contains a small number of repeated values. Choosing the right values and giving them short codes can produce significant file compression.

Colour palette

Dictionary coding can be used to compress colour images. The dictionary gives a short code to each colour used in the image. A dictionary of colours is called a colour palette.

PRACTICE QUESTION

2 A colour photo of a seascape is stored at resolution 1200 × 1800 pixels. The image format uses RGB colour codes. The colour photo has 32 different shades of blue and white, with no other colours.

a The image uses RGB colour. Give the number of bytes used to store each pixel.

b Give the number of bits used to store each pixel.

c A colour palette is created. It has a short code for each colour used in the image. State how many codes are needed.

d State how many bits are needed to store each colour code in the dictionary.

You will answer more questions about this photo example in Topic 6.5.

REMEMBER: Dictionary compression will work on any file where data values are repeated. They need not occur in large blocks as with RLE. A data dictionary stores the most common values used in the file. Each value is given a short code. The code is used instead of the value throughout the file.

6.5 The effect of compression

Measure compression

To measure the effect of compression:

- calculate the size of the original file
- calculate the size of the compressed file
- find the difference between the two values.

On this page you will see a worked example. This shows you how to work out the effect of compression. The worked example uses the DNA text file example from the previous spread.

The practice questions give you the chance to apply your understanding. The practice questions refer to the seascape photo example from the previous spread.

WORKED EXAMPLE

A file contained four letters (A, C, G, and T) organised in a varied sequence of 200 000 characters. The letters were stored using ASCII. Calculate the size of the uncompressed file.

First, find the original file size of this data.

The sequence consists of 200 000 characters – each stored using a 1 byte ASCII code.

$$200\,000 \times 1 \text{ byte} = 200 \text{ kB}$$

The uncompressed file is 200 kB.

Calculate the reduction in file size that would be achieved by compression using dictionary coding.

There are only four data values in the file. To store four codes, two bits are needed.

Value	Dictionary code
A	00
C	01
G	10
T	11

There are 200 000 characters in the file.

$$200\,000 \times 2 \text{ bits} = 400\,000 \text{ bits}$$

Divide by eight to find the number of bytes.

$$\frac{400\,000}{8} = 50\,000$$

The compressed file is 50 kB.

Compression has reduced the file size by 150 kB.

PRACTICE QUESTION

1 A colour photo of a seascape is stored at resolution 1200 × 1800 pixels. The image format uses RGB colour codes (24-bit colour).

- **a** Calculate how many pixels are in the image.
- **b** Give the number of bits used to store each pixel.

- **c** Calculate how many bits are used to store the whole image.
- **d** Convert this to a file size and express the result using a suitable digital unit.
- **e** The image is compressed using a colour palette of 32 values. The codes are the numbers 0–31.
  - **i** State how many bits are used to store each colour code.
  - **ii** Calculate how many bits are needed to store the value of every pixel in the image.
  - **iii** Convert this to a file size using a suitable digital unit.

Expressing the results of compression

You can express the results of compression as an absolute value. Subtract the size of the compressed file from the size of the original file.

original file size – compressed file size = reduction in file size

Or you can express the results as a percentage. Divide the reduction in file size by the original file size and multiply by 100 to express the reduction in percentage terms.

$$\frac{\text{reduction in file size}}{\text{original file size}} \times 100 = \text{percentage reduction}$$

WORKED EXAMPLE

Dictionary coding reduced the size of a file from 1.2 MB to 300 kB. Calculate:

- **a** the absolute reduction in file size
- **b** the percentage reduction in file size.

To find the absolute reduction in file size, subtract the smaller file size from the larger file size. (Tip: Make sure they are in the same units.)

1200 kB – 300 kB = 900 kB

The absolute reduction in file size is 900 kB.

To find the percentage reduction in file size, divide the reduction by the original file size, then multiply by 100.

$$\frac{900}{1200} \times 100 = 75\%$$

PRACTICE QUESTION

3 In question 1, you found the size of an uncompressed file. In question 2, you found the size of the same file after dictionary compression.

- **a** State the absolute reduction in file size.
- **b** Express the reduction in file size as a percentage (to the nearest whole number).

REMEMBER: To calculate the effect of compression on file size:

1. calculate the original file
2. calculate the compressed file size
3. subtract the compressed size from original size to give the change in file size.

6.6 Find the size of a compressed file

Different methods of compression

On this spread you will apply what you have learnt about compression to find the size of a compressed file.

Case study

The worked examples relate to a file with the following features.

- The image contains 2720 pixels.
- The file format uses 24-bit colour coding (RGB colour).
- The file contains pixels of the following five colours: black, white, red, blue, and green.
- The pixels occur in continuous blocks of (on average) 16 pixels.

This file is compressed using a number of different methods and the results are compared.

WORKED EXAMPLE

Calculate the size of the file before compression.

In 24-bit colour, each pixel is stored using 24 bits.

Multiply the number of bits per pixel by the number of pixels in the image.

$$24 \times 2720 = 65\,280 \text{ bits}$$

Divide the number of bits by eight to give the number of bytes

$$\frac{65\,280}{8} = 8160 \text{ bytes}$$

Expressing this value using suitable units; the original file is 8.16 kB.

The file is compressed using RLE. Estimate the size of the compressed file.

RLE changes the file into a series of blocks. Each block starts with the data value and then a number that represents the number of occurrences.

In this case the blocks average 16 pixels. There are 2720 pixels altogether.

$$\frac{2720}{16} = 170$$

So there are 170 blocks of continuous colour. In RLE each block is represented by a pair of values:

- the data value – in this case a colour code (3 bytes)
- the frequency (1 byte).

The total is 4 bytes per block. So the file is represented by 170 blocks of 4 bytes.

$$170 \times 4 = 680 \text{ bytes}$$

The file is compressed using a colour palette (dictionary compression). Estimate the size of the compressed file.

The colour palette is a dictionary that stores each colour using a short binary code. The image has five colours so the dictionary needs five different codes.

Colour	Binary code
black	000
white	001
red	010
blue	011
green	100

Each pixel is stored using a 3-bit binary code. Multiply the number of bits per pixel by the number of pixels in the image.

$3 \times 2720 = 8160\,\text{bits}$

Divide the number of bits by eight to give the number of bytes.

$\frac{8160}{8} = 1020\,\text{bytes} = 1.02\,\text{kB}$

Summary of results

File	File size	Reduction in kB	% Reduction
Original file	8.16 kB	n/a	n/a
RLE compression	680 bytes	7.48	91.7
Dictionary compression	1.02 kB	7.14	87.5

In this case, RLE compression is more effective. This is due to the nature of the data. Remember that many compression systems combine both dictionary and RLE compression.

PRACTICE QUESTION

1 Here are some facts about a data file.

The file contains 13 200 characters, stored using 8-bit ASCII.

The file contains the characters A, B, C, D, and E only.

On average, the letters appear in blocks of 33 repeated characters.

The file is compressed using RLE compression and dictionary compression.

Complete the following table to show the size of the file before and after RLE and dictionary compression.

File	File size	Reduction	% Reduction
Original file			
RLE compression			
Dictionary compression			

7 HUFFMAN COMPRESSION (AQA ONLY)

7.1 The Huffman algorithm

Compression algorithm

An algorithm sets out the steps needed to solve a problem. One type of algorithm is a compression algorithm. A compression algorithm sets out the steps for compressing a file.

Huffman compression is an algorithm for dictionary compression.

The Huffman algorithm

You have seen how dictionary coding works to compress a file – Huffman compression is a way of building the dictionary. The Huffman algorithm builds a dictionary to store the different values used in the file.

Huffman compression also creates the codes for the data values.

- The values that occur most often get the shortest codes.
- The values that occur least often get longer codes.

By using the shortest codes for the most common values the Huffman algorithm achieves the greatest possible compression.

WORKED EXAMPLE

Huffman tree

The Huffman algorithm goes through a data file and organises all the values into a tree shape. The values that occur most often appear at the top of the tree. The values that appear less often appear lower down the tree.

Here is an example. An image file has 272 pixels. The pixels are red, blue, green, black, and white. The file has:

- 107 red pixels
- 75 blue pixels
- 60 green pixels
- 20 white pixels
- 10 black pixels.

The Huffman tree is shown in Figure 1. The most common value is red, and this appears at the top of the tree. The least common values are black and white. These appear at the bottom of the tree.

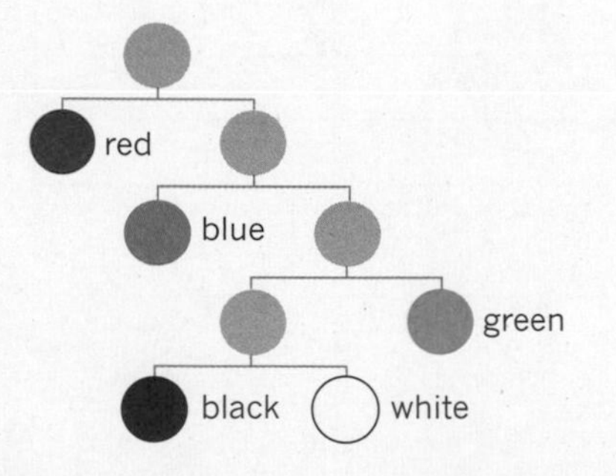

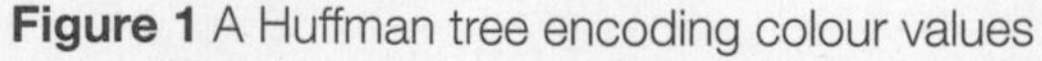
Figure 1 A Huffman tree encoding colour values

Find Huffman code

To find a Huffman code for a particular value:

- trace the path from the top of the tree to the value you want
- every time you take a left branch write a 0
- every time you take a right branch write a 1
- stop when you reach the value you want.

The list of 1s and 0s makes a binary code. To get to a value high on the tree is a short path. The binary code is shorter. To get to a value low on the tree is a longer path. The binary code is longer.

WORKED EXAMPLE

Identify the code for red using the example Huffman tree.

1 Start at the top.

2 Take the left branch (write 0).

3 You have found 'red'.

In this file, the Huffman code for red is 0. It is a frequent value. It has a short code.

Identify the code for black using the example Huffman tree.

1 Start at the top.

2 Turn right (write 1).

3 Turn right (write 1).

4 Turn left (write 0).

5 Turn left (write 0).

6 You have found 'black'.

In this file, the Huffman code for black is 1100. It is a less frequent value, with a longer code.

PRACTICE QUESTION

1 Use Figure 1 to find the codes for:

  a blue

  b green

  c white

2 Figure 2 shows a Huffman tree that encodes six values represented by the letters X, T, Z, Q, H, and M.

  Use this tree to identify the Huffman codes for the following values.

  a X   **b** Z   **c** Q   **d** H

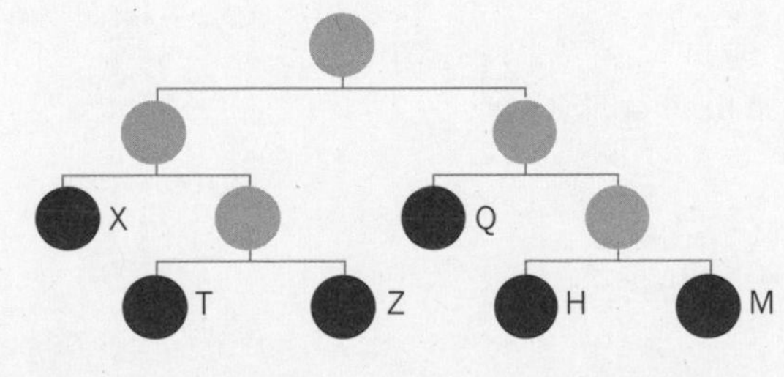

Figure 2 A Huffman tree encoding six letters

NOTE: A letter always has the same ASCII code in every file where ASCII is used.

A letter does not always have the same Huffman code in every file that uses Huffman coding.

Huffman codes are based on the frequency of a value in a data file. If the letter has a different frequency in another file it will have a different Huffman code.

REMEMBER: To make the Huffman code for a data value, start at the top of the tree. Move down the tree towards the value you want to encode. When you turn left write 0. When you turn right write 1.

7.2 Interpreting a Huffman tree

Find values that match codes

The Huffman tree lets us find the value that corresponds to any Huffman code.

As you have learnt, a Huffman code is a series of 1s and 0s. The list of 1s and 0s makes a binary code. To get to a value high on the tree is a short path. The binary code is shorter. To get to a value low on the tree is a longer path. The binary code is longer.

To turn the code into a data value:

- start at the top of the tree
- count through the bits of the Huffman code from left to right
- if the bit is 0, branch left
- if the bit is 1, branch right
- when there are no more bits, you have reached the value you want.

WORKED EXAMPLE

The Huffman tree in Figure 1 contains the letters A–E. Identify the letter represented by the code 101.

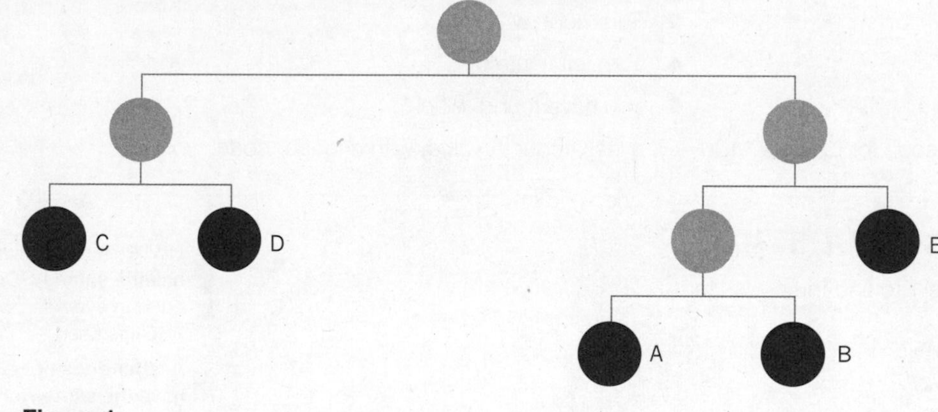

Figure 1

Start at the top of the tree. Use the same rule as before. Turn left for 0, turn right for 1.

- The first bit is 1, so take the right branch.
- The second bit is 0, so take the left branch.
- The final bit is 1, so take the right branch.
- You have reached the letter B.

So the Huffman code 101 stands for the letter B in this tree.

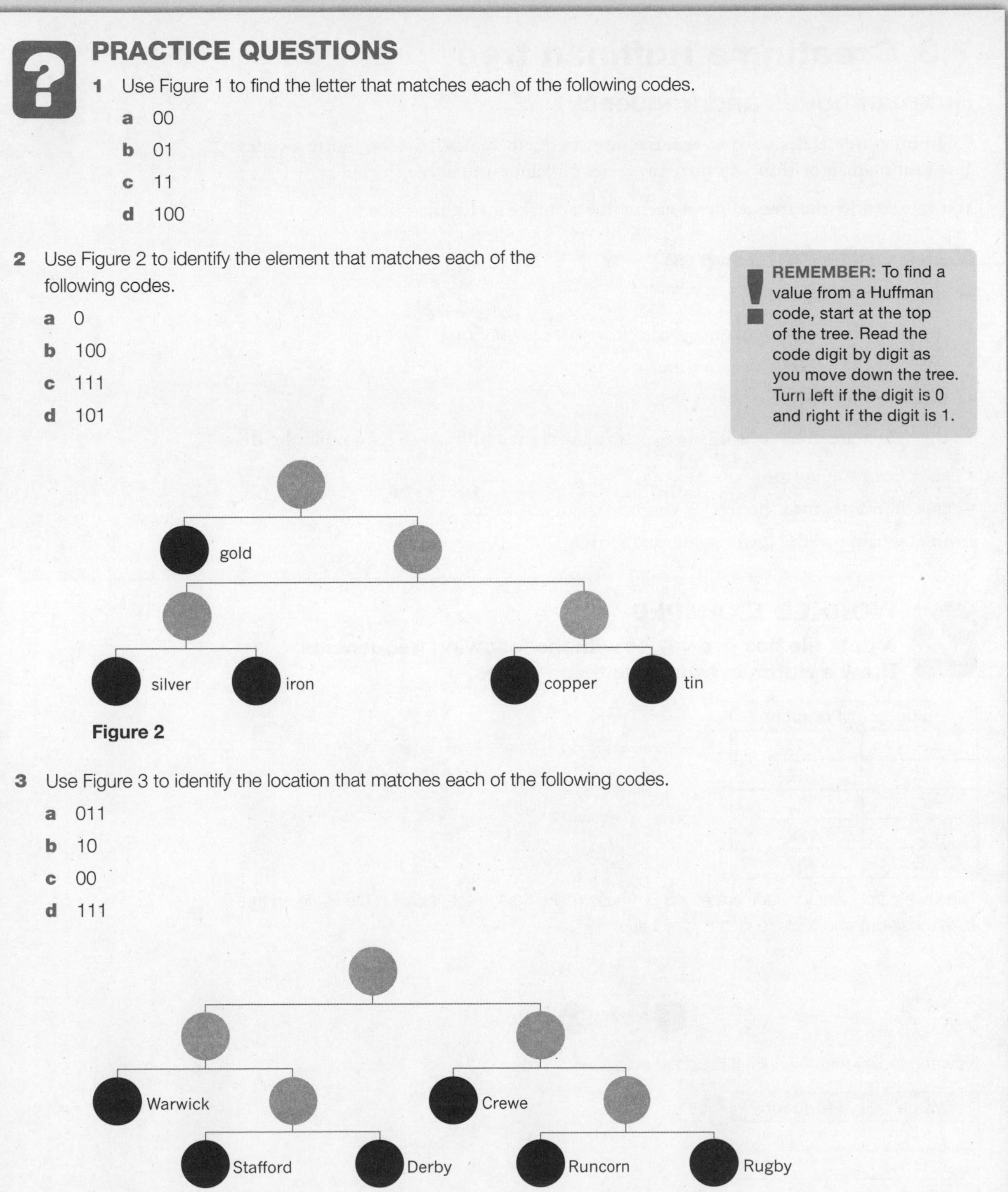

PRACTICE QUESTIONS

1 Use Figure 1 to find the letter that matches each of the following codes.

a 00

b 01

c 11

d 100

2 Use Figure 2 to identify the element that matches each of the following codes.

a 0

b 100

c 111

d 101

REMEMBER: To find a value from a Huffman code, start at the top of the tree. Read the code digit by digit as you move down the tree. Turn left if the digit is 0 and right if the digit is 1.

Figure 2

3 Use Figure 3 to identify the location that matches each of the following codes.

a 011

b 10

c 00

d 111

Figure 3

7.3 Creating a Huffman tree

Huffman codes and frequency

A Huffman tree is designed so that the most frequent values have the shortest codes. The Huffman algorithm will turn any series of values into a tree shape.

You must know the frequency of each value to make a Huffman tree.

Make codes into a tree

To make a Huffman tree:

- put the values into frequency order, lowest frequency first
- take out the values two at a time
- take the least frequent values first.

As the values are taken out in pairs you can build the Huffman tree. To build the tree:

- start from the bottom
- put the values into the tree as you take them out of the list
- in any pair, put the lower value on the right.

WORKED EXAMPLE

A data file has five values with the following frequencies. Draw a Huffman tree using these values.

Value	Frequency
A	14
B	19
C	21
D	22
E	37

The lowest frequency values are A and B. These make the lowest values in the Huffman tree. Take these out and replace with a joint value.

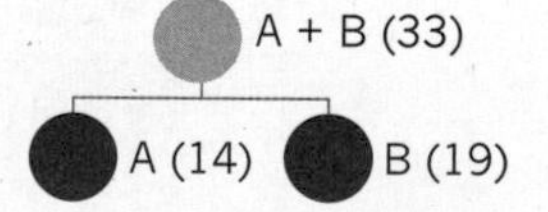

Now the table looks like this. It is sorted in frequency order.

Value	Frequency
C	21
D	22
A + B	33
E	37

The lowest frequency values are C and D. Take these out and replace with a joint value.

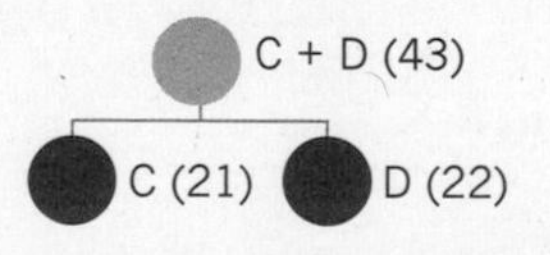

Now the table looks like this.

Value	Frequency
A + B	33
E	37
C + D	43

The lowest values are the A + B pair and the value E. Combine these values, with the lowest to the left.

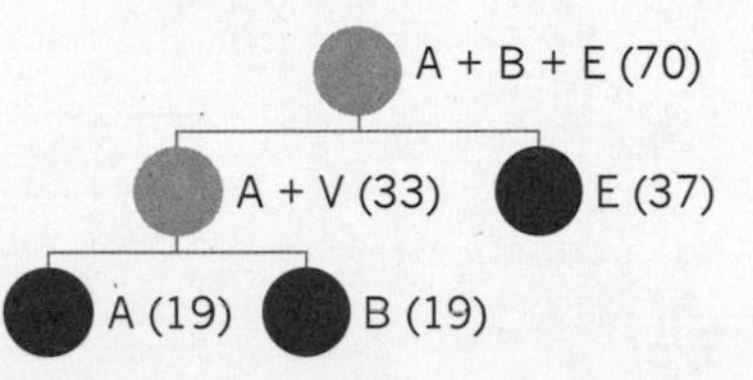

The table now looks like this.

Value	Frequency
C + D	43
A + B + E	70

Complete the tree by combining the final values.

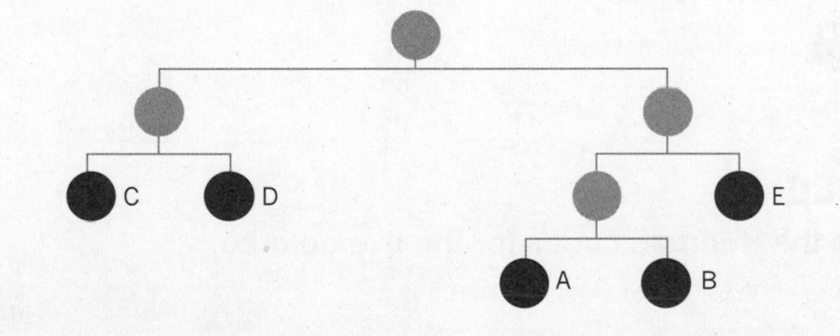

PRACTICE QUESTION

1 Give the Huffman codes for the five values in the tree shown above.

2 Use the following frequency values to draw a Huffman tree for the five colours.

Colour	Frequency
black	200
white	150
blue	60
green	50
yellow	30

7.4 Find the size of a Huffman file

Case study

The worked examples relate to a file with the following features.

- The image contains 2720 pixels.
- The file format uses 24-bit colour coding (RGB colour).
- The file contains pixels of the following five colours: black, white, red, blue, and green.
- The number of pixels of each colour is shown in the next table.

Colour	Number of pixels
red	1070
blue	750
green	600
black	200
white	100

The original file size is 8.16 kB (see Topic 6.6 for the calculation).

The file is compressed using the Huffman tree in Figure 1.

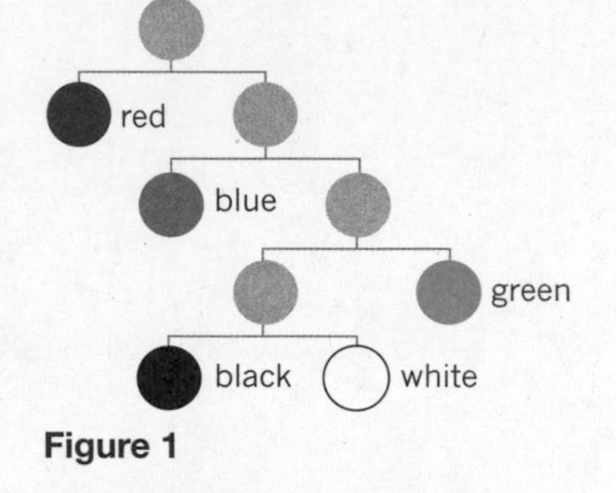

Figure 1

WORKED EXAMPLE

Look at Figure 1. Give the Huffman codes for the five colours.

You have learnt how to find the code for a value in a Huffman tree (Topic 7.2). The table on the right shows the code for each colour.

Colour	Binary code
red	0
blue	10
green	111
black	1100
white	1101

Calculate the size of the compressed file.

Each colour is coded using a short code. Red is the most common colour so it has the shortest code.

Colour	Binary code	Bits per pixel
red	0	1
blue	10	2
green	111	3
black	1100	4
white	1101	4

The next table shows the number of pixels of each colour. Multiply the number of bits per colour by the number of pixels.

Colour	Bits per pixel	Number of pixels	Total
red	1	1070	$1070 \times 1 = 1070$
blue	2	750	$750 \times 2 = 1500$
green	3	600	$60 \times 3 = 1800$
black	4	200	$200 \times 4 = 800$
white	4	100	$100 \times 4 = 400$

Add up the total number of bits used:

$$1070 + 1500 + 1800 + 800 + 400 = 5570 \text{ bits}$$

Divide by eight to give the total number of bytes:

$$\frac{5570}{8} = 696.25 \text{ – round this up to 697 bytes or 0.7 kB}$$

Conclusion

The original file was 8.16 kB. After Huffman compression the file is 0.7 kB. Subtract the reduced file size from the original file size to get the reduction in file size.

- 8.16 – 0.7 = 7.46
- The reduction in file size was 7.46 kB.

Divide reduction by original size to get percentage reduction.

- $\frac{7.46}{8.16} = 0.914$
- The percentage reduction is 91.4%

PRACTICE QUESTION

1 A text file:

- contains 13400 characters.
- contains the characters A, B, C, D, and E only
- is stored using 8-bit ASCII.

The letters occur in the following frequencies.

Letter	Frequency
A	3000
B	900
C	2000
D	500
E	7000

The data is encoded in the following Huffman tree.

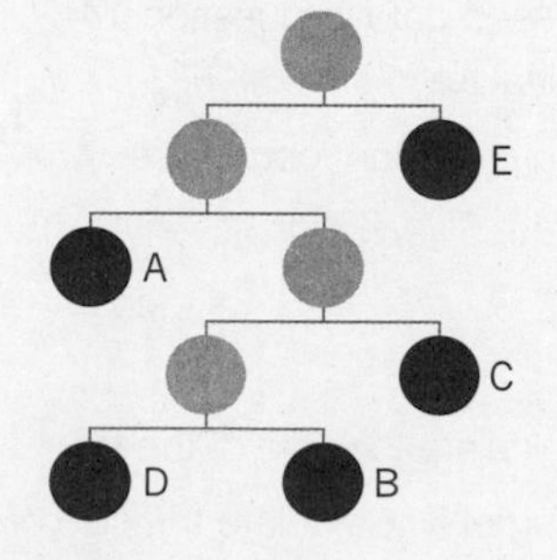

a Give the Huffman code for each letter.

b Calculate the effect of Huffman compression on the file, as an absolute reduction and a percentage reduction.

SUMMARY QUESTIONS FOR CHAPTERS 4–7

1 Interpret the following bitstream as a monochrome image of dimensions 6 × 6 pixels. It has been divided into nibbles to make it easier to read.

`1000 0110 1101 0100 1001 0010 1011 0110 0001`

2 A monochrome image has resolution 120 × 340 pixels. Express the size of the image in bits and bytes. Convert to any suitable units.

3 State how many colour codes are available in 8-bit colour.

4 Show how the colour white would be represented in RGB colour. Use decimal (base 10) values.

5 Show how the colour white would be expressed in RGB colour using three binary numbers.

6 Calculate the file size of an image that uses 16-bit colour and has 120 × 240 pixels. Express the answer using suitable units.

7 A 10-second audio recording uses a sample rate of 11.5 kHz. Calculate how many digital samples the file contains in total.

8 An audio recording has a sample resolution of 16 bits. It has a sample rate of 44.1 kHz. Calculate the bit rate.

9 An audio recording has a bit rate of 350 000. The recording lasts 2 minutes. Calculate the size of the file. Express the answer using a suitable digital unit.

10 An image file with 420 pixels was stored using 32-bit colour. Calculate the file size.

11 An image file with 420 pixels was compressed by converting 32-bit colour to 8-bit colour. Calculate the reduction in file size.

12 State whether the compression in the previous question is an example of lossless or lossy compression. Explain your answer.

13 The following data string is stored using 8-bit ASCII.
Give the total storage requirement.

XXXXXXXYYYYYYYYYYYYXXXXXXXXXXXXXYYYYYYYYXXXXXXXXXXYYY

14 Show how the data string from question 13 could be expressed as a series of value/frequency pairs using RLE compression.

15 Calculate the storage requirement of the RLE-compressed data in question 14. How much of a reduction is this compared to the original file?

16 The following 8-bit ASCII string was compressed using dictionary compression. Calculate the storage requirement of the data before and after compression.

ZXYYYZYZZYZXXZYXZXXZ

17 A 30 × 60 pixel image contains 45 different colours.

- **a** The colours are stored using 24-bit true colour. Calculate the size of the file.
- **b** The image is compressed using dictionary encoding only. Calculate the size of the compressed file.
- **c** The image contains colours in blocks of four pixels. It is compressed using run-length encoding only. Calculate the size of the compressed file.
- **d** Compare the reduction in file size achieved by the two forms of compression given in parts b and c. You may use absolute values rather than percentages. Identify which form of compression is more effective.

HUFFMAN CODING

An image contains pixels of five colours stored using 32-bit colour. Pixels occur in the following frequencies.

Colour	Frequency
light green	1200
cyan	800
moss green	400
teal	1500
sea green	900

18 Calculate the total file size of the image before compression.

A Huffman tree is created from this data set

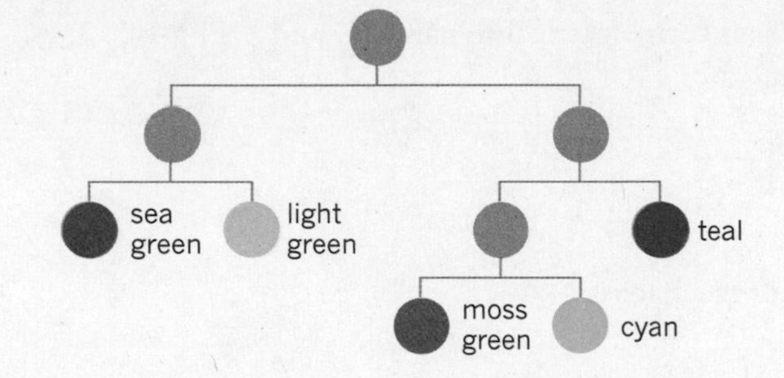

19 Create a table showing the Huffman code for each character in the data set.

20 Calculate the total file size after Huffman compression.

21 Calculate the percentage reduction that has been achieved compared to the uncompressed file.

Give the answer to the nearest whole number.

8.1 The NOT gate

Electricity inside the computer

All data inside the computer is represented using on/off electrical circuits. ON and OFF circuits are used to represent the 1s and 0s of binary numbers. All data values are represented using binary numbers.

The data values inside a computer are not fixed. They can change. That is how the computer processes data. Electricity flows through the circuits inside the computer processor. Some signals are turned on and some are turned off.

Logic gates

Logic gates are small electronic circuits that can turn electrical signals ON and OFF. In this book you will learn about three logic gates:

- the NOT gate
- the AND gate
- the OR gate.

Each gate has a different effect on the flow of electricity.

Inputs and outputs

Every logic gate can have one or two wires going into the gate. These are called the input signals.

Every logic gate has exactly one wire coming out of the gate. This is called the output signal.

Input and output signals can be in one of two states.

- The wire is carrying electricity. We call this an ON signal.
- The wire is not carrying electricity. We call this an OFF signal.

ON and OFF signals

As well as calling them ON and OFF signals we can use True/False values.

- An ON signal can be represented by the value True or the number 1.
- An OFF signal can be represented by the value False or the number 0.

PRACTICE QUESTION

1 The output of a logic gate is a binary signal. State which answer best explains why we call the output a binary signal.

  a The signal can be in two states (ON or OFF).

  b The signal has 8 bits (1 byte).

  c There are two inputs to the gate.

  d The signal is represented using a binary number.

The NOT gate

The NOT gate is an electronic circuit inside the computer. Electricity flows into the NOT gate. Electricity flows out of the NOT gate.

The NOT gate changes the electrical signal.

- If the electrical signal going into the NOT gate is ON, the signal coming out is OFF.
- If the signal going into the NOT gate is OFF, the signal coming out is ON.
- In other words, the NOT gate reverses the signal. The output signal is the opposite of the input signal.

PRACTICE QUESTION

2 **a** State how many inputs are there to a NOT gate.

b State how many outputs a NOT gate produces.

c The input to a NOT gate is an OFF signal. Give the output of the gate.

d The input to a NOT gate is an ON signal. Give the output of the gate.

Representing the NOT gate

Computer scientists sometimes need to draw a diagram of the circuits inside a computer. Circuits are made of logic gates joined together. In real life gates are very small. Instead of drawing the real-life logic gate, computer scientists use a symbol.

Here is the symbol for the NOT gate.

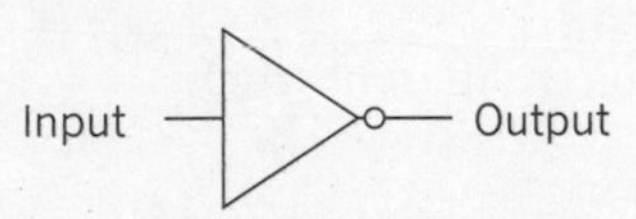

Notice the small circle in the drawing. That is an important part of the symbol.

PRACTICE QUESTION

3 Without looking at the picture in this book, practise drawing the NOT gate symbol. Mark the input and output wires.

REMEMBER: Truth gates are electronic circuits. They have inputs and outputs. Inputs can be ON or OFF. Outputs can be ON or OFF.
The NOT gate has one input and one output.
The NOT gate reverses the value of the Input.
It turns ON into OFF.
It turns OFF into ON.

8.2 The AND gate

The AND gate is an electronic circuit inside the computer. Electricity flows into the AND gate. Electricity flows out of the AND gate.

The AND gate has two inputs. The AND gate produces one output.

- If both inputs to an AND gate are ON, the output is ON.
- In all other cases the output of the AND gate is OFF.

PRACTICE QUESTION

1 a State how many inputs there are to the AND gate.

b Give the number of outputs produced by the AND gate.

c The inputs to an AND gate are both OFF. State the output.

d The inputs to an AND gate are both ON. Give the output.

e One input to an AND gate is ON. The other input is OFF. State the output.

Representing the AND gate

Each logic gate has its own symbol.
The symbol for the AND gate is shown on the right.

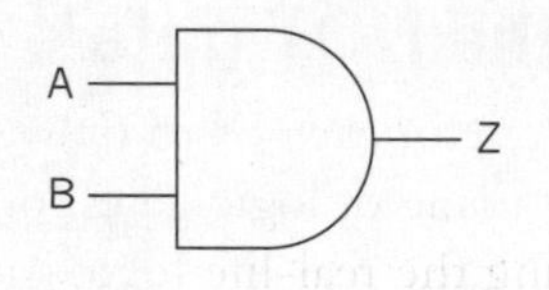

Notice that the front of the AND gate is curved and the back is straight. These are important features of the drawing.

Labelling the inputs and outputs

The AND gate has two inputs. The two inputs must have different labels.
The inputs to a logic gate are often labelled A and B.

The output from a logic gate is labelled with a different letter. Often, it is a letter that is far from A and B in the alphabet. In this example the letter Z is used, but other letters can be used.

PRACTICE QUESTION

2 Without looking at the picture in this book, practise drawing the AND gate symbol. Mark the inputs and outputs using suitable letters.

Input states

The AND gate has two inputs. We can label them A and B. The AND gate has four possible input states.

- A is OFF and B is OFF.
- A is OFF and B is ON.
- A is ON and B is OFF.
- A is ON and B is ON.

This table shows the four input states of the AND gate.

Input A	Input B
OFF	OFF
OFF	ON
ON	OFF
ON	ON

PRACTICE QUESTION

3 a Draw a table showing the four input states of the AND gate. Use the binary digits 1 and 0 to represent ON and OFF signals.

b Identify which binary numbers are made by these digits.

Everyday logic

The NOT gate and the AND gate operate like the words 'not' and 'and' in everyday speech. Later you will see how these gates can be used inside the computer to solve real-world logic problems.

The word 'NOT'

The NOT gate operates like the word 'not' in everyday speech. It changes a True/False value to its opposite.

- 'I am going to a party'.
- 'I am NOT going to a party'.

Adding the word NOT has reversed the True/False value.

- If the first sentence is true, the second sentence is false.
- If the first sentence is false, the second sentence is true.

The word 'AND'

The AND gate operates like the word 'and' in everyday speech. A sentence with 'and' in it is only true if both part of the sentence are true.

For example, a rollercoaster operator might ask, 'Are you older than 15 AND taller than 1.5 m?'

The word AND separates two parts of the question:

- are you older than 15?
- are you taller than 1.5 m?

The answer to the whole question is only True if both parts are True.

REMEMBER: The AND gate has two inputs and one output.
The output of the AND gate is ON only if both inputs are ON.

8.3 The OR gate

The OR gate is an electronic circuit inside the computer. The OR gate has two inputs. The OR gate produces one output.

- If both inputs to the OR gate are OFF, the output is OFF.
- In all other cases the output of the OR gate is ON.

PRACTICE QUESTION

1 a Give the number of inputs to the OR gate.

b State how many outputs the OR gate produces.

c The inputs to an OR gate are both OFF. Give the output.

d The inputs to an OR gate are both ON. Give the output.

e One input to an OR gate is ON. The other input is OFF. Give the output.

Representing the OR gate

Each logic gate has its own symbol. Here is the symbol for the OR gate. The inputs and outputs are labelled with letters.

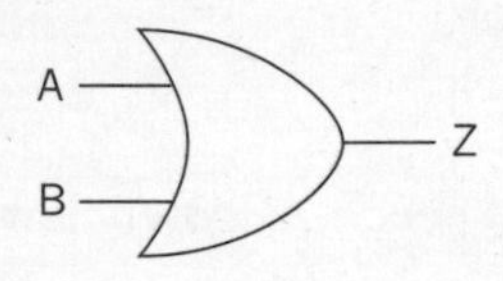

Notice that the front of the OR gate is pointed and the back is curved. These are important features of the drawing. They are how you recognise that this is the OR gate instead of the AND gate.

Input states

The OR gate has two inputs. The OR gate has four possible input states. These are the same as the four input states of the AND gate.

PRACTICE QUESTION

2 a Without looking at the pictures in this book, practise drawing the AND and OR gate symbols. Make sure it is easy to tell the difference between the two symbols.

b Draw a table showing the four input states of the OR gate. Use the binary digits 1 and 0 to represent ON and OFF signals.

Everyday logic

Similarities to everyday speech

The OR gate operates like the word 'or' in everyday speech. It joins two statements together. For example, a person might ask 'Will you grow any flowers OR vegetables?'

The word 'OR' joins together the two parts of the question:

- will you grow any flowers?
- will you grow any vegetables?

The overall answer is true if at least ONE of the answers to the parts of the question is true. The overall answer is also true if both parts are true.

This is like the way the OR gate operates inside the computer. The output is True if at least one of the inputs is True.

Differences from everyday speech

In everyday speech we sometimes use the word 'or' in a different way. For example, a person might say:

'I have to study OR I will fail the exam.'

What the person means is that one or the other statement will be true, but not both. This is a different way of using the word OR. Inside the computer the OR gate is not used like this. If both inputs are True, the output is True.

REMEMBER: The OR gate has two inputs and one output. The output of the OR gate is OFF if both inputs are OFF. In all other cases the output is ON.

PRACTICE QUESTION

3 Draw a line from each logic gate to its description. Label each gate with its correct name.

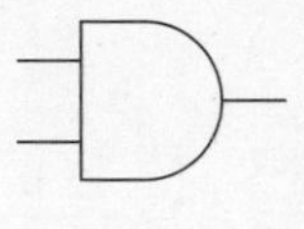

Reverses the electrical charge from ON to OFF and vice-versa

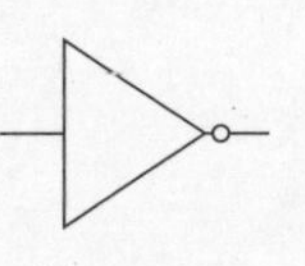

Takes two inputs. Will output an electrical charge unless both inputs are OFF.

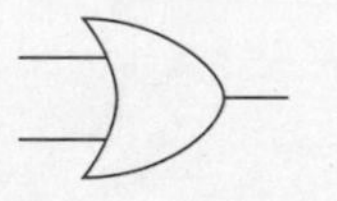

Takes two inputs. Will only produce a positive output if both inputs carry an electrical charge.

8.4 Truth tables

Input states

Each input to a logic gate can be ON or OFF. Different settings or combinations of inputs are called input states.

A truth table lists all the possible inputs states of a logic gate. The truth table also shows the output that goes with each input state.

The NOT gate

A truth table shows all the possible inputs to a logic gate. The NOT gate has only one input signal and it can be in two different states.

Input signal
ON
OFF

The truth table also shows the output of the NOT gate. It is always the opposite of the input.

Input signal	Output signal
ON	OFF
OFF	ON

Remember we can show ON and OFF signals as True/False values or as 1s and 0s. We can label inputs and outputs using letters.

Input A	Output X
1	0
0	1

PRACTICE QUESTION

1 Without looking at the example in this book, draw:

- **a** the truth table for the NOT gate using ON/OFF values
- **b** the truth table for the NOT gate using True/False values
- **c** the truth table for the NOT gate using 1s and 0s

The AND gate

The AND gate has two inputs. They are typically labelled A and B. Each input can be ON or OFF. We can show these using the binary digits 1 and 0.

The AND gate can be in four different states.

A	B
0	0
0	1
1	0
1	1

The truth table shows the output of the AND gate for each input state. In this example the output is labelled Q.

The output is only ON if both inputs are ON.

A	B	Q
0	0	0
0	1	0
1	0	0
1	1	1

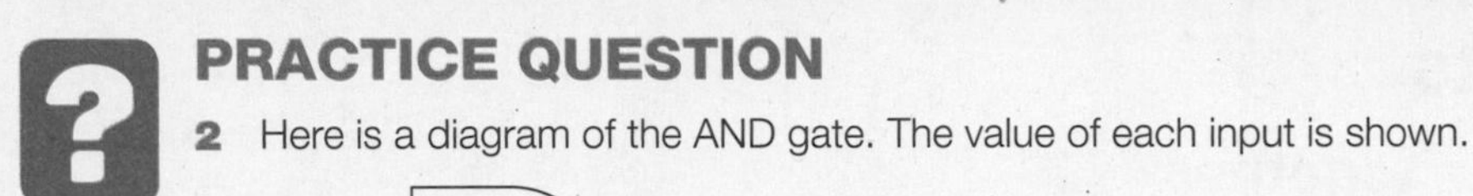

PRACTICE QUESTION

2 Here is a diagram of the AND gate. The value of each input is shown.

ON

OFF

a Copy this diagram and add a label to show the value of the output in this case.

b Draw a second diagram showing the AND gate with the output ON. State the inputs.

The OR gate

The OR gate has two inputs. Each input can be ON or OFF. The OR gate can be in four different states. These are the same as the input states of the AND gate.

A	B
0	0
0	1
1	0
1	1

This truth table shows the output of the OR gate. In this example the output is labelled Q.

A	B	Q
0	0	0
0	1	1
1	0	1
1	1	1

The output is ON if one or both inputs are ON.

PRACTICE QUESTION

3 These gate symbols and truth tables have been mixed up. Draw a line from each logic gate to its matching truth table.

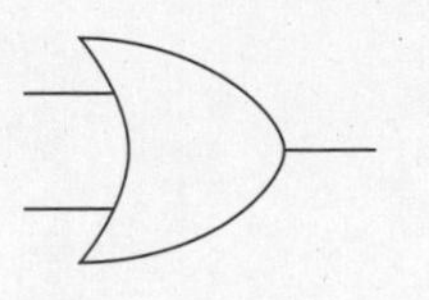

A	B	Q
0	0	0
0	1	0
1	0	0
1	1	1

A	B
0	0
0	1

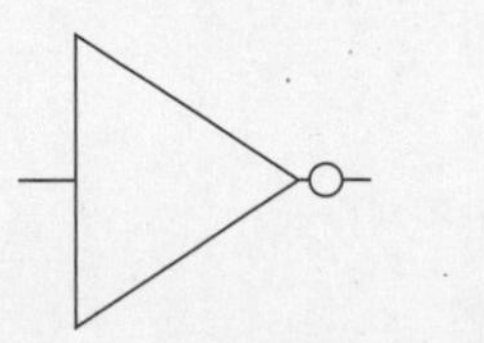

A	B	Z
0	0	0
0	1	1
1	0	1
1	1	1

REMEMBER: A truth table for a logic gate has a column for each input to the gate. It also has a column for the output. Each row of the table shows a different possible state for the gate. Overall the truth table shows all possible inputs to the gate, and the resulting output in each case.

8.5 Logic circuits

Connecting logic gates together

Logic gates can be wired together. The output of one gate becomes the input of the next gate. A series of logic gates wired together is called a logic circuit.

A truth table can be made to match the logic circuit. The truth table will show:

- all the possible states of the logic circuit
- the output of the logic circuit in each case.

WORKED EXAMPLE

Draw a truth table that shows all possible states of the following logic circuit. Use the truth table to find the inputs that will produce a True output (output signal on).

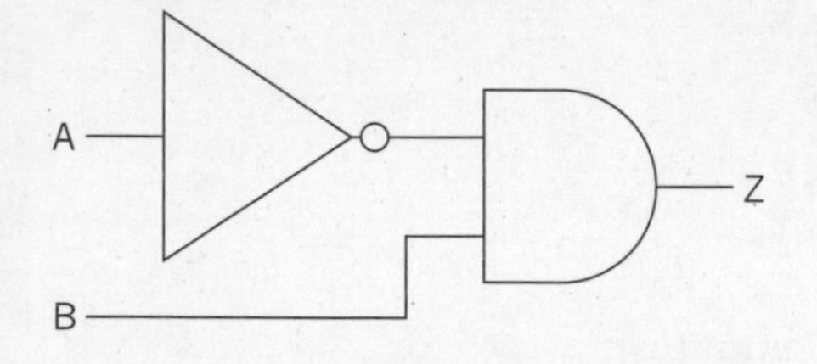

This logic circuit contains two logic gates: a NOT gate and an AND gate. The output of the NOT gate is wired into the AND gate.

A good first step when analysing a logic gate is to look for any unlabelled outputs and give them a label. The example circuit has one unlabelled output – the output of the NOT gate. Here it is given the label C.

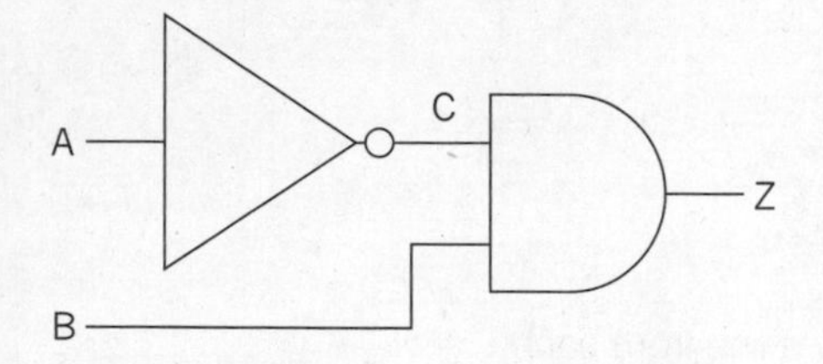

Now draw a truth table to match this logic circuit. The truth table should have a column to match every labelled input and output in the circuit.

The next step is to fill in the inputs. The logic circuit has two inputs: A and B. This means it has four possible input states. This truth table shows all inputs to the logic circuit. ON and OFF signals are shown as 1s and 0s.

A	B	C	Z
0	0		
0	1		
1	0		
1	1		

To complete the truth table, fill in the output columns. Begin with output C.

C is the output of a logic gate. The gate is a NOT gate. The input to the NOT gate is A. So C has the value NOT A. In every line, the value of C is the reverse of A. Complete the C column by filling in values that are the reverse of A.

Z is the output of a logic gate. The gate is an AND gate. The inputs to the AND gate are B and C. Complete the Z column with 1 when B and C are both 1. In all other cases, fill in the value 0.

A	B	C = NOT A	Z = B AND C
0	0	1	0
0	1	1	1
1	0	0	0
1	1	0	0

We can see from this truth table that only one combination of inputs will produce the output value 1. This is the second row of the table, where A is 0 and B is 1.

PRACTICE QUESTIONS

1 **a** Draw the following logic circuit, labelling any unlabelled outputs.

b Draw a truth table that shows all possible states of the logic circuit.

c Use the truth table to find the inputs that will produce a True output (output signal 1).

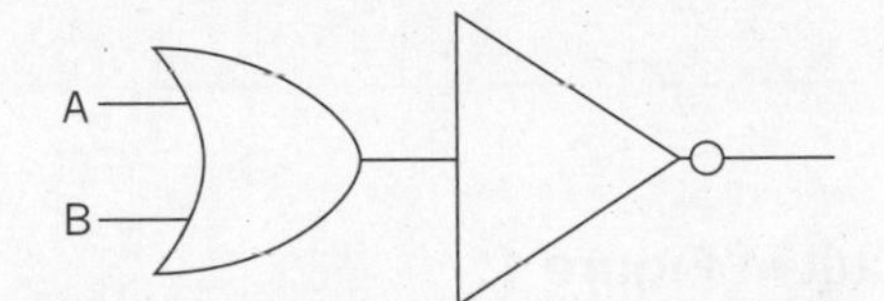

2 **a** Draw the following logic circuit, labelling any unlabelled outputs.

b Draw a truth table that shows all possible states of the logic circuit.

c Use the truth table to find what inputs will produce a True output (output signal on).

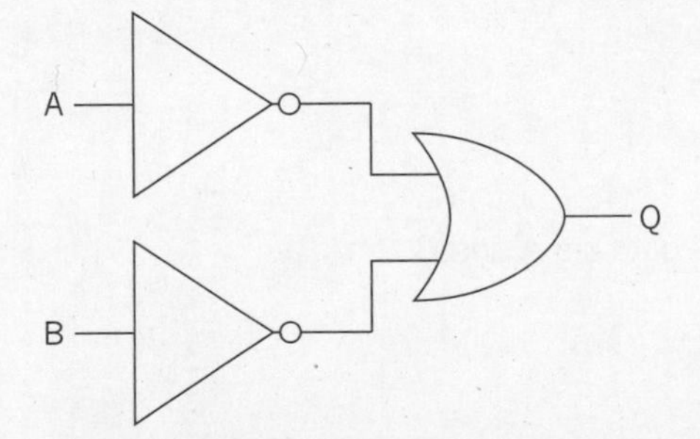

REMEMBER: To create a logic gate for a circuit, label every input and output in the circuit. Draw a table with a column for each input and output. Make a row for each possible combination of inputs. If there are two inputs this will match the first four binary numbers: 00, 01, 10, 11. Complete each output column. What type of gate produces the output? For each row of the table check the inputs to that gate and fill in the correct output value.

8.6 Logic circuits with shared inputs

Shared inputs

Logic circuits are made by wiring logic gates together. A logic circuit diagram shows how the logic gates are wired together. The output of one gate becomes the input of another gate.

In some circuits the connecting wire is divided. The same signal is sent to two different gates.

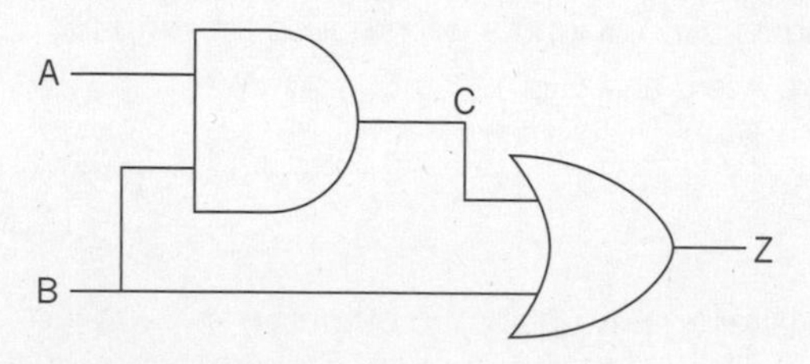

Figure 1 A logic circuit with shared inputs

In Figure 1, the wire that carries input signal B is split.
Signal B is sent to both gates in the circuit.

WORKED EXAMPLE

Draw a truth table to match the logic circuit in Figure 1.

In this example all the outputs have already been labelled. Draw a truth table with a column for each value in the table, and fill in the input values A and B.

A	B	C	Z
0	0		
0	1		
1	0		
1	1		

The output C is the output of an AND gate. The two inputs to the AND gate are A and B. Complete this column.

A	B	C = A AND B	Z
0	0	0	
0	1	0	
1	0	0	
1	1	1	

The output Z is the output of an OR gate. The two inputs to the OR gate are C and B.
Complete this column.

A	B	C = A AND B	Z = C OR B
0	0	0	0
0	1	0	1
1	0	0	0
1	1	1	1

PRACTICE QUESTIONS

1 a Draw the following logic circuit, labelling any unlabelled outputs.

b Draw a truth table that shows all possible input states of the logic circuit.

c Use the truth table to identify which inputs will produce an output of 1.

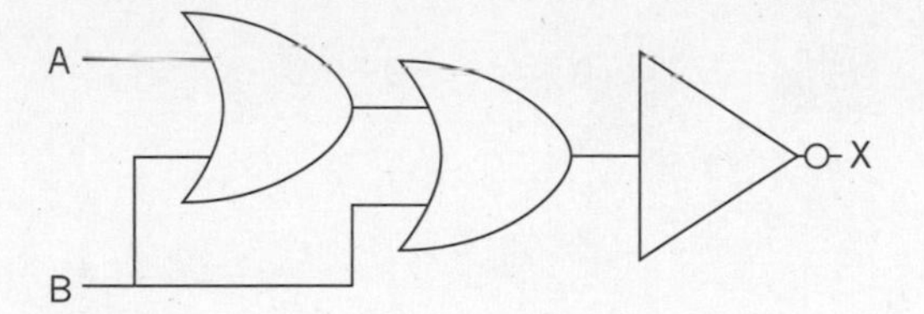

2 a Draw the following logic circuit, labelling any unlabelled outputs

b Draw a completed truth table for the logic circuit.

c Use the truth table to find which inputs will produce the output 1.

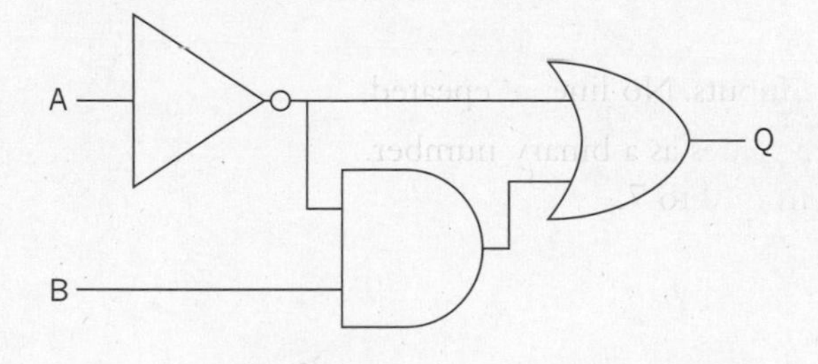

8.7 Logic circuits with three inputs

More than two inputs

The number of states that a logic circuit can be in will depend on how many inputs it has.

- A logic circuit with one input can be in two different states. The input can be ON or OFF.
- A logic circuit with two inputs can be in four different states. You have created several truth tables with four rows showing these four states.
- A logic circuit with three inputs can be in eight different states.

Where a logic circuit has three inputs they are usually labelled A, B, and C.

A three-input truth table

The next table shows the eight states of a logic circuit with three inputs.

A	B	C
0	0	0
0	0	1
0	1	0
0	1	1
1	0	0
1	0	1
1	1	0
1	1	1

This table shows all possible combinations of the three inputs. No line is repeated.

Read each line of the table in turn and write the three values as a binary number. The sequence matches the first eight binary numbers from 0 to 7.

000

001

010

011

100

101

110

111

This is a good way to remember all possible states of a circuit with three inputs.

WORKED EXAMPLE

Draw a truth table to match the logic circuit diagram shown.

All inputs and outputs are labelled. Draw a table with a column for each label. There should be five columns.

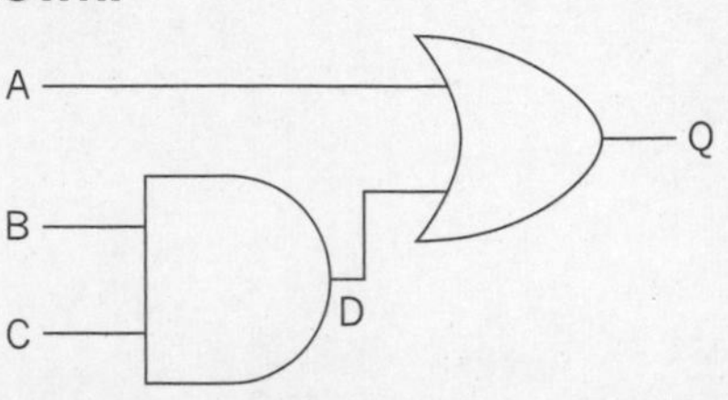

There are three inputs so there are eight possible input states. The table should have eight rows.

Complete the A, B, and C columns to show all possible input states.

A	B	C	D	Q
0	0	0		
0	0	1		
0	1	0		
0	1	1		
1	0	0		
1	0	1		
1	1	0		
1	1	1		

Now fill in the remaining columns.

D is the output of an AND gate. The inputs to the gate are B and C. So D has the value B AND C. D will have the value 1 if both B and C are 1.

Q is the output of an OR gate. The inputs to the gate are A and D. So Q has the value A OR D. Q has the value 1 if either A or D or both are 1.

A	B	C	D = B AND C	Q = A OR D
0	0	0	0	0
0	0	1	0	0
0	1	0	0	0
0	1	1	1	1
1	0	0	0	1
1	0	1	0	1
1	1	0	0	1
1	1	1	1	1

PRACTICE QUESTIONS

1 **a** Draw the following logic circuit, labelling any unlabelled outputs.

b Draw a truth table that shows all possible input states of the logic circuit.

c Complete the truth table and use it to identify which inputs will produce an output of 1.

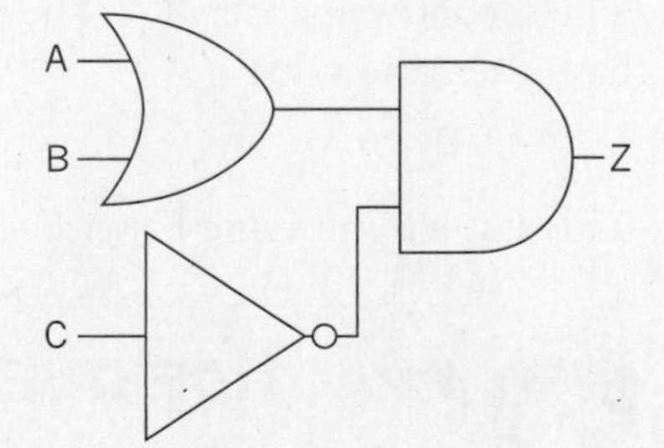

2 **a** Draw the following logic circuit, labelling any unlabelled outputs.

b Draw a truth table for this logic circuit.

c Use the truth table to find which inputs will produce an output of 0.

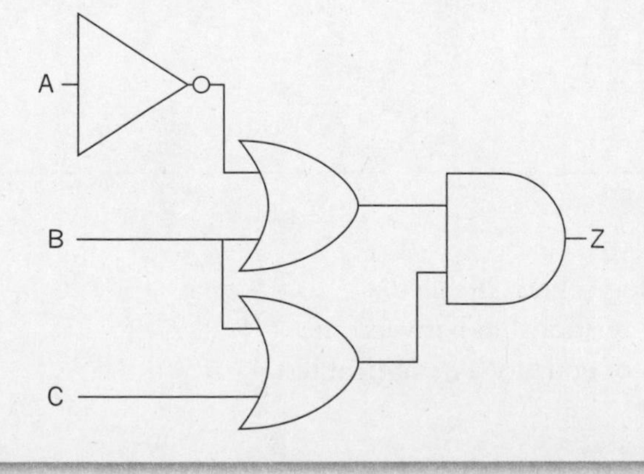

REMEMBER: If there are three inputs to a logic circuit the circuit can have eight possible input states. These correspond to the first eight binary numbers: 000, 001, 010, 011, 100, 101, 110, 111

9 LOGICAL EXPRESSIONS

9.1 Logical tests

What is an expression?

In computer science an expression is a set of symbols that represents a value. Several different expressions can have the same value. For example, all these expressions have the same value:

$12 + 3$

5×3

$21 - 6$

They all have the value 15.

Compare two expressions

A logical test compares two expressions. The comparison is either true or false. Here is an example of a logical test:

$10 = 12$

This test compares the expressions 10 and 12. The test says '10 is equal to 12'. The logical test has the value False (because 10 is not equal to 12).

Calculated values

Instead of numbers the two expressions that are compared could be calculations. To complete the logical test:

- work out the value of each expression
- compare the two values.

For example:

$10 = 5 \times 2$

The second value is 5×2. The value of this expression is ten. So the two expressions have the same value.

$12 = 5 \times 3$

This test has the value False (because 12 is not equal to 5×3).

PRACTICE QUESTION

1 For each logical test state whether it has the value True or False.

a $120 = 121$

b $100 = 10 \times 10$

c $(1 + 9) = (3 + 7)$

d $(3 \times 7) = (2 \times 4)$

Relational operator

The symbol that goes between the two expressions in a logical test is called a relational operator. The relational operator represents the relationship between the two expressions. The equals sign represents a relationship of equality. The logical test has the value True if the two expressions are equal.

This relational operator can be written as = or as ==. For example:

10 == 31 − 21

This test is has the value True (because 10 is equal to 31 − 21).

Not equal

A logical test can also check if two expressions are not equal. The relational operator that checks if two expressions are NOT equal can be written as != or as ≠. You may see either operator used.

- The test has the value True if the expressions are NOT EQUAL.
- The test has the value False if the expressions are EQUAL.

For example:

10 != 19

This test has the value True (because 10 is not equal to 19).

12 ≠ (3 × 4)

This test has the value False (because 12 is equal to 3 × 4).

PRACTICE QUESTION

2 For each logical test state whether it has the value True or False.

a 120 != (6 × 20)

b 34 ≠ (7 × 5)

c (99 + 1) ≠ (25 × 4)

d 130 != (12 × 10)

Non-numeric values

The examples on this page compare two numerical expressions. But logical tests can also compare other expressions, such as characters, text strings, variables, and other data values. The two values must be identical for the expressions to be equal.

Here are two examples. Both these logical tests have the value True.

"A" == "A"

"A" != "a"

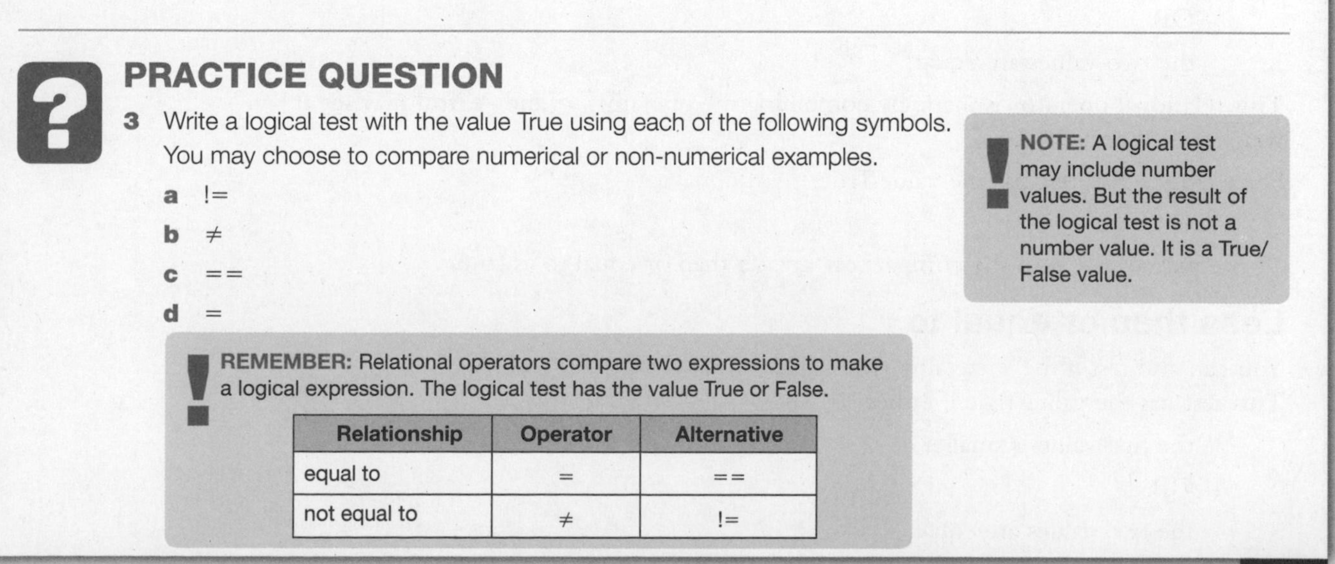

PRACTICE QUESTION

3 Write a logical test with the value True using each of the following symbols. You may choose to compare numerical or non-numerical examples.

a !=

b ≠

c ==

d =

NOTE: A logical test may include number values. But the result of the logical test is not a number value. It is a True/False value.

REMEMBER: Relational operators compare two expressions to make a logical expression. The logical test has the value True or False.

Relationship	Operator	Alternative
equal to	=	==
not equal to	≠	!=

9.2 Relational operators

The > and < operators

Logical tests compare two values. One way to compare two values is to see if one value is greater than the other.

$10 > 12$ means '10 is greater than 12'

$10 < 12$ means '10 is less than 12'

The symbols $>$ and $<$ are relational operators. Relational operators compare two values. If the relational operator correctly describes the relationship between the two expressions then the logical test has the value True.

Relationship	Operator
greater than	$>$
less than	$<$

WORKED EXAMPLE

Evaluate the logical test 15 > (3 × 5).

The relational operator is >, which means 'greater than'.

- The expression before the relational operator is the number 15.
- The expression after the relational operator is 3 × 5, which has the value 15.

So the test asks 'Is 15 greater than 15?' The answer is no (15 is not greater than 15).
The logical test has the value False.

PRACTICE QUESTION

2 For each logical test state whether it has the value True or False.

a 120 < 121

b 100 > 1000

c (1 + 9) < (3 + 7)

d (3 × 7) > (2 × 4)

Greater than or equal to

Another way to compare two values is to see if the first is 'greater than or equal to' the second. This test has the value True if either:

the first value is larger

OR

the two values are equal.

This relational operator is made by combining the operators $>$ and $=$. You may see it written as $\geq$ or as $>=$.

For example, this test has the value True.

15 >= (3 × 5)

15 is equal to 3×5 so the comparison 'greater than or equal to' is true.

Less than or equal to

You can also test if the first value is 'less than or equal to' the second.
This test has the value True if either:

the first value is smaller

OR

the two values are equal.

This relational operator is made by combining the operators < and =. You may see it written as ≤ or as <=.

For example, this test has the value False.

15 ≤ (2.5 × 5)

15 is greater than 2.5 × 5 so the comparison 'less than or equal to' is false.

PRACTICE QUESTIONS

2 For each logical test state whether it has the value True or False.

a 120 ≤ 121

b 100 >= 1000

c (1 + 9) <= (3 + 7)

d (3 × 7) ≥ (2 × 4)

3 Give a logical test with the value False using each of the following symbols.

a > **b** <

c >= **d** <=

STRETCH YOURSELF!

Comparing non-numerical expressions

The greater than and less than operators can be used with non-numerical expressions. In these cases the computer will compare the ASCII (or Unicode) values of the characters.

ASCII codes are sequential. Upper case characters have ASCII values from 65 to 91. Lower case ASCII characters have ASCII values from 97 to 123.

You do not have to remember the ASCII codes to compare characters.

- Upper case characters are 'less than' lower case characters.
- A character earlier in the alphabet is 'less than' a character later in the alphabet.

Give the result of the logical test "A" > "T".

The letter A is earlier in the alphabet than the letter T. That means its ASCII value is lower.

The logical test uses the operator >, which means 'greater than'.

The comparison is incorrect so the test has the value False.

PRACTICE QUESTION

4 Give the results of the following logical tests.

a "D" < "h"

b "dog" == "Dog"

c "Q" <= "q"

d "ABC" > "abc"

REMEMBER:

Relationship	Operator	Alternative
greater than or equal to	≥	>=
less than or equal to	≤	<=

9.3 Boolean expressions

Boolean values

'Boolean' values are the values True and False.

Boolean expressions are expressions that can have the value True or False.

Boolean operators

Boolean expressions are made by joining Boolean values together with Boolean operators.

The Boolean operators you will learn are:

NOT

AND

OR

There are others, but they are outside the scope of this book.

You have already learnt the rules that go with these operators. They match the three logic gates you learnt about in Chapter 8.

NOT

The NOT operator can go in front of any Boolean value. It reverses the value of the expression. Here is a simple example:

NOT True

This expression has the value False. The NOT operator has reversed the value.

You can put a NOT operator in front of any expression that has a True/False value. It will reverse the value.

WORKED EXAMPLE

Give the value of the following Boolean expression.

NOT (4 > 5)

First, work out the value of the logical test inside brackets. The logical test 4 > 5 has the value False. Putting NOT in front reverses the value. So the Boolean expression has the value True.

PRACTICE QUESTION

1 Give the value of the following Boolean expressions.

- **a** NOT (9 == 10)
- **b** NOT (3 ≠ 2 + 8)
- **c** NOT ("Q" <= "q")
- **d** NOT (5 < 2 × 3)

AND

The AND operator can join any two Boolean values. This creates a larger Boolean expression. The expression is true if both Boolean values are True. In all other cases the value of the Boolean expression is False.

WORKED EXAMPLE

If A is False and B is True, give the value of the following Boolean expression.

A AND B

AND is used to join two Boolean values to make a larger Boolean expression. The larger expression is only True if both the joined values are True.

You have been told that A is False. This means the larger expression is False.

PRACTICE QUESTION

2 Give the value of the following Boolean expressions.

- **a** (3 < 4) AND (5 < 6)
- **b** (7 = 8) AND (9 > 3)
- **c** ("a" = "A") AND True
- **d** 2 × 2 = 4 AND 99 < 100

NOTE: Boolean expressions are named after an English mathematician called George Boole. He lived in Victorian times. He set out the rules of logic in a methodical way that we still use today.

OR

The OR operator can join any two Boolean values. This creates a larger Boolean expression.

- The larger expression is True if one or both of the values is True.
- The larger expression is False if both values are False.

WORKED EXAMPLE

If A is False and B is True, give the value of the following Boolean expression.

A OR B

OR is used to join two Boolean values to make a larger Boolean expression. The larger expression is True if either one of the joined values is True.

You have been told that B is True. This means the larger expression is True.

PRACTICE QUESTION

3 Give the value of the following Boolean expressions.

- **a** (3 < 4) OR (5 < 6)
- **b** (7 = 8) OR (9 > 3)
- **c** ("a" = "A") OR False
- **d** 2 × 2 = 4 OR 99 < 100

REMEMBER: Boolean operators join Boolean values to make Boolean expressions.
To calculate the truth value: First work out the True/False values of each part. Then combine the values using the operators to find the value of the larger expression.

9.4 Complex Boolean expressions

Boolean values

In the examples on this page Boolean expressions will be made using letters A, B, and C. These letters will have the following values.

- A is True.
- B is True.
- C is False.

The same expressions could be made using any True/False values, such as logical tests, or letters with any other truth value. But for simplicity in these examples we will use the three letters shown. The letters can stand for any expression.

More than one operator

Longer Boolean expressions can be made using more than one operator. Here is an example:

(A AND B) OR C

WORKED EXAMPLE

State the value of the following Boolean expression.

(A AND B) OR C

To find the value of the whole expression, find the values of the two parts.

- Find the value of (A AND B). A Boolean expression using AND is only True if the values on either side of the operator are both True, so (A AND B) has the value True.
- Find the Value of C. You know C is False.

So the expression reads:

True OR False

An expression made with OR is True if at least one of the expressions on either side of the operator is True. That is the case here. So the whole expression has the value:

True

PRACTICE QUESTION

1 Give the value of the following Boolean expressions.

a (A OR B) OR C

b (A AND B) AND C

c (A OR B) AND C

d NOT C

Brackets first

A Boolean expression with more than one operator may include brackets. Here is an example:

A AND (B OR C)

If a Boolean expression includes brackets, then find the Truth value of the expression in brackets first.

WORKED EXAMPLE

State the value of the following Boolean expression.

A AND (B OR C)

Work out the expression in brackets first. This is (B OR C). An expression with OR is True if either of the values are True. As B is True, (B OR C) is True.

A AND (True)

An expression with AND is True if both values are True. You know A is True. So the whole expression is:

True

PRACTICE QUESTION

2 A is True, B is True, and C is False. Give the value of the following Boolean expressions.

a B AND (C OR A)

b NOT (A AND C)

c (NOT A) AND (NOT C)

d NOT (B OR C)

Order of precedence

If there are no brackets in an expression find the value of expressions in this order:

NOT

AND

OR

WORKED EXAMPLE

State the value of the following Boolean expression.

A AND NOT B OR C

There are three operators in this expression: AND, OR, and NOT. Order of precedence puts NOT first.

The NOT operator reverses truth value. B is True, so NOT B is False. Substituting this value, the expression reads:

A AND False OR C

There are two operators remaining in this expression: AND and OR. Order of precedence puts AND first.

An AND expression is only True if the values on either side are true. So (A AND False) is False. Substituting this value, the expression reads:

False OR C

We know that C is False so the expression reads:

False OR False

An expression with OR is true if either of the joined expressions is True. In this case both the joined expressions are False, so the overall value is:

False

PRACTICE QUESTION

3 Give the value of the following Boolean expressions.

a A OR C AND B

b A AND NOT C

c NOT C OR B

d NOT B OR C AND A

REMEMBER: To calculate the value of an expression with multiple operators, calculate the value of any part that is in brackets. Then use the order of precedence: NOT then AND then OR.

9.5 Truth tables and Boolean expressions

Truth table

On the previous page you were given truth values for A, B, and C. From this you worked out the value of a Boolean expression made with A, B, and C. There is another way to set out truth values. That is to look at all possible truth values of an expression. You can do this using a truth table.

You already made truth tables for logic circuits. A truth table sets out all possible states. It shows all the possible truth values of A, B, and C. You can make a truth table for a Boolean expression in exactly the same way.

WORKED EXAMPLE

Draw a truth table for the following Boolean expression.

(A OR B) AND C

There are three inputs to the expression: A, B, and C. Make a table with columns for each input value. Fill in all possible values. Reading across the table you should see all the binary numbers, counting from 000 to 111.

A	B	C
0	0	0
0	0	1
0	1	0
0	1	1
1	0	0
1	0	1
1	1	0
1	1	1

Now you will add columns for each part of the expression.

Remember from the previous page that you work out the part in brackets first. This is (A OR B). Add a column to the table with the heading A OR B. A OR B is True if A is True or B is True.

A	B	C	A OR B
0	0	0	0
0	0	1	0
0	1	0	1
0	1	1	1
1	0	0	1
1	0	1	1
1	1	0	1
1	1	1	1

To complete the truth table, add the final expression. The operator is AND. This operator joins (A OR B) to C. The statement is true if (A OR B) is True and C is also True.

Fill in these values:

A	B	C	A OR B	(A OR B) AND C
0	0	0	0	0
0	0	1	0	0
0	1	0	1	0
0	1	1	1	1
1	0	0	1	0
1	0	1	1	1
1	1	0	1	0
1	1	1	1	1

The table now shows all possible truth values for the expression.

PRACTICE QUESTION

1 Draw truth tables for the following Boolean expressions.

a (A AND B) OR C

b (A OR B) OR C

c (A AND B) AND NOT C

d NOT (A AND B) AND C

Brackets and order of precedence

You build up the truth table one expression at a time. Pick which expression to evaluate using the order of precedence that you learnt in Topic 9.4.

As you evaluate an expression, add a column to the truth table. When you have evaluated all the expressions, the truth table is completed.

PRACTICE QUESTIONS

2 Draw truth tables for the following Boolean expressions.

a (NOT B) OR C

b NOT (B OR C)

c (NOT A) AND (NOT C)

d (NOT A) OR (B AND C)

3 Draw truth tables for the following Boolean expressions.

a A OR C AND B

b A OR B AND NOT C

c A AND NOT B OR C

d NOT A AND B OR C

REMEMBER: To create a truth table from an expression, make a column for each variable used in the expression. Fill these columns to show every possible combination of 1s and 0s.
Add columns for each part of the expression. Use the order of precedence.
Add 1s and 0s in each column by working out the True/False values. Use the rules for Boolean operators that you have already learnt.

9.6 Boolean algebra

Terms for Boolean operators

In this chapter you have used the Boolean operators:

NOT

AND

OR

You may see these Boolean operators given longer names.

Operator	Name
NOT	negation
AND	conjunction
OR	disjunction

PRACTICE QUESTION

1 Match the Boolean terms with the correct Boolean expressions.

disjunction	A AND B
conjunction	NOT A
negation	B OR C

Symbols for Boolean operators

You may see symbols used instead of the words AND, OR, NOT.

Operator	Symbol
NOT	¬
AND	∧
OR	∨

These operators are used just like the words and have an identical meaning.

WORKED EXAMPLE

Create a truth table for the following Boolean expression. Give the values of A and B needed to give a True value for the whole expression.

$A \land \neg B$

The expression has two input values, A and B. So the truth table has four rows.

A	B
0	0
0	1
1	0
1	1

The symbol ¬ stands for NOT and the symbol ∧ stands for AND.

NOT has precedence over AND. So evaluate ¬B first.

A	B	¬B
0	0	1
0	1	0
1	0	1
1	1	0

The complete the expression by evaluating ∧, which represents AND.

A	B	¬B	A ∧ ¬B
0	0	1	0
0	1	0	0
1	0	1	1
1	1	0	0

The expression is only True if A is True and B is False.

PRACTICE QUESTIONS

2 Draw truth tables to match the following Boolean expressions.

a ¬A ∨ B

b A ∨ (B ∨ C)

c A ∧ ¬(B ∧ C)

d ¬(A ∧ ¬B)

3 Complete the following table. Fill in all the rows so that each shows:

- a logic gate
- the equivalent Boolean expression
- the descriptive term
- the Boolean algebra symbol.

Logic gate	Boolean expression	Descriptive term	Boolean algebra
	NOT		
			∧
		Disjunction	

10 PROBLEM SOLVING

10.1 Logical problems

Real-life Boolean expressions

In Chapter 9 you learnt how logical expressions can be represented using Boolean operators. In these examples the letters A, B, and C were used to stand for truth values. A, B, and C can have the values True or False.

Boolean expressions can be applied to real-life problems. Instead of the letters 'A', 'B', and 'C' the expression contains sentences that can be True or False. Logic will find the combination of values that produces a True result.

WORKED EXAMPLE

A bank vault can be opened by entering the correct key pad code or by a card swipe plus retinal scan. Give this as a Boolean expression.

The bank vault can be opened in two ways:

- by entering the correct key pad code
- by a card swipe plus retinal scan.

If either of these actions is True then 'open the vault' is True. The first action can be True, or the second can be True, or both. The operator that corresponds to this is the OR operator.

key pad OR (card swipe plus retinal scan)

The expression in brackets contains two actions. These are card swipe and retinal scan. In this case, both the card swipe and the retinal scan must be True to unlock the vault. The operator that corresponds to this is the AND operator.

key pad OR (card swipe AND retinal scan)

You could also write this as a simplified Boolean expression:

A OR (B AND C)

where A, B, and C stand for the different ways of opening the vault.

PRACTICE QUESTION

1 Give Boolean expressions to express these real-world examples.

a Your user name must begin with a capital letter and contain at least one digit.

b An alarm is sounded if a door is opened and a code is not entered.

c There are two ways to pass a course: get over 80% in an exam, or get 60% and then complete a practical test.

d To pick up a parcel you must wait two days, then show ID or sign a form.

WORKED EXAMPLE

A bank vault can be opened with a key pad code or by a card swipe plus retinal scan. Draw a truth table to represent all possible states of the vault.

If you have not done so already, you must turn the example into a Boolean expression. This was done in the previous worked example.

A OR (B AND C)

A stands for key pad, B stands for card swipe, and C stands for retinal scan.
You can then make a truth table for the expression using the skills you have already learnt earlier in this book. First, fill in all possible input values.

A (key pad)	B (card swipe)	C (retinal scan)
0	0	0
0	0	1
0	1	0
0	1	1
1	0	0
1	0	1
1	1	0
1	1	1

Then complete the outputs. Take the part that is in brackets first. Then evaluate the whole expression.

A	B	C	B AND C	A OR (B AND C)
0	0	0	0	0
0	0	1	0	0
0	1	0	0	0
0	1	1	1	1
1	0	0	0	1
1	0	1	0	1
1	1	0	0	1
1	1	1	1	1

PRACTICE QUESTION

2 Draw truth tables to match each of these real-world examples.

a Your user name must begin with a capital letter and contain at least one digit.

b An alarm is sounded if a door is opened and a code is not entered.

c There are two ways to pass a course: get over 80% in an exam, or get 60% and then complete a practical test.

d To pick up a parcel you must wait two days, then show ID or sign a form.

10.2 Logic circuits and Boolean expressions

What logic circuits are for

There are many similarities between logic circuits and Boolean expressions. Computers have logic circuits so they can work out the value of Boolean expressions. They find the truth values very quickly and accurately using the flow of electricity.

A logic circuit can be made to match any Boolean expression. The logic circuit can then be used to solve any real-world problem that matches the Boolean expression.

Match operators to logic gates

A logic circuit can be built to match any Boolean expression. Choose a logic gate that matches the Boolean operator in the expression. Label the inputs to the gate to match the values used in the expression.

WORKED EXAMPLE

Draw a logic circuit diagram to match the following Boolean expression.

B OR C

The Boolean operator is OR so draw the OR gate. The two expressions joined by the operator are B and C. So label the inputs to the gate B and C. In this example the output has been given the value D – but it can be any letter.

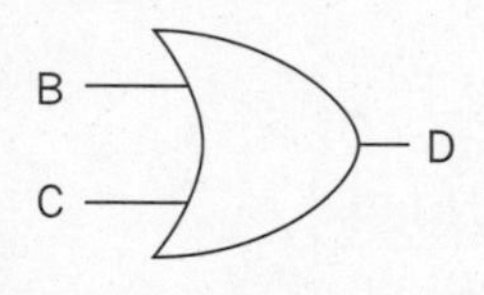

More than one operator

Convert the Boolean operators into logic gates. If there is more than one operator then use the rules of precedence to decide which operator to convert first.

Remember brackets first, then NOT, AND, OR.

PRACTICE QUESTION

1 Draw logic circuits to match the following Boolean expressions.

a A AND B

b NOT C

WORKED EXAMPLE

Draw a logic circuit diagram to match the following Boolean expression.

NOT (B OR C)

Convert the expression in brackets first.

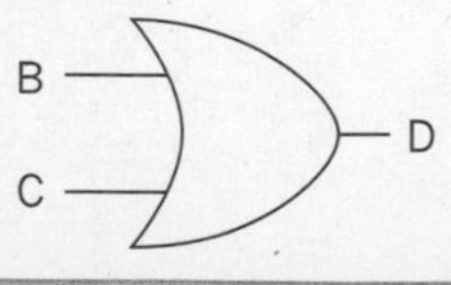

The remaining operator is NOT. This reverses the truth value of the OR gate. So put the output from the OR gate through a NOT gate. The result is called E.

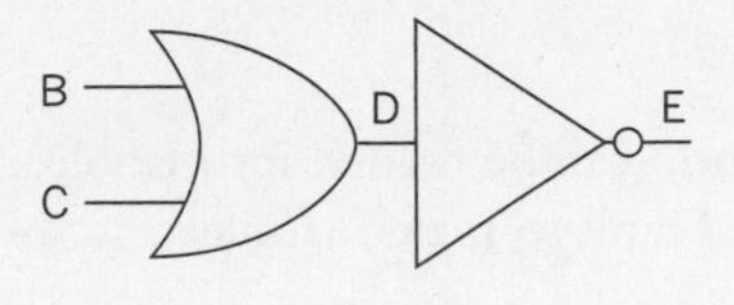

Draw a logic circuit diagram to match the following Boolean expression.

A OR (B AND C)

Convert the expression in brackets first.

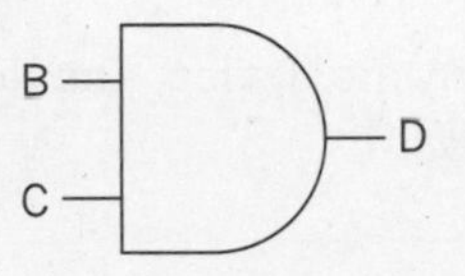

The remaining operator is OR. So put the output of this gate into an OR gate.

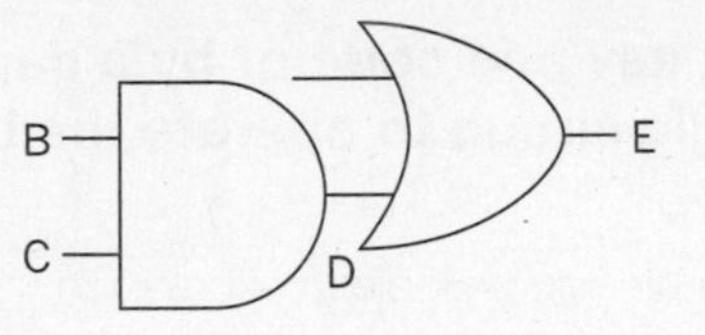

The OR operator joins (B AND C) to A. So the other input to the OR gate is A.

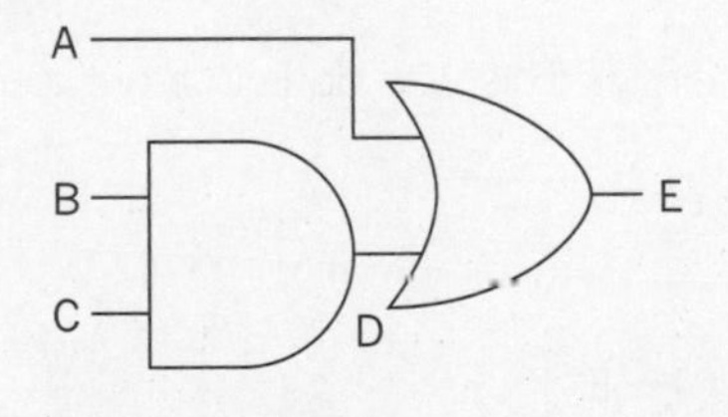

PRACTICE QUESTIONS

2 Draw logic circuits to match the following Boolean expressions.

- **a** (NOT A) AND B
- **b** NOT (A AND B)
- **c** (NOT C) OR B
- **d** A OR (NOT B)

3 Draw logic circuits to match the following Boolean expressions.

- **a** A AND (B OR C)
- **b** A AND B OR C
- **c** NOT A AND (B OR C)
- **d** NOT A AND B OR C

10.3 Logical problem solving

Using logic to solve problems

You have seen how logical problems in the real-world can be turned into Boolean expressions. You have seen how logic circuits can be built to match Boolean expressions.

This is how computers can be built to solve problems. The computer will direct electricity through a series of logic gates that match a real-world problem. The computer will quickly and accurately find the combination of inputs that produces the required output.

For example, a computer can be used to control an electronic device. The device will operate when the correct combination of inputs is received.

WORKED EXAMPLE

A bank vault can be opened with a key pad code or by a card swipe plus retinal scan. Draw a logic circuit to operate the bank vault opening switch.

If you have not done so already, you must turn the example into a Boolean expression. This was done in a previous worked example (Topic 10.1).

A OR (B AND C)

Then you must turn the Boolean expression into a logic circuit. This was done in a previous worked example (Topic 10.2).

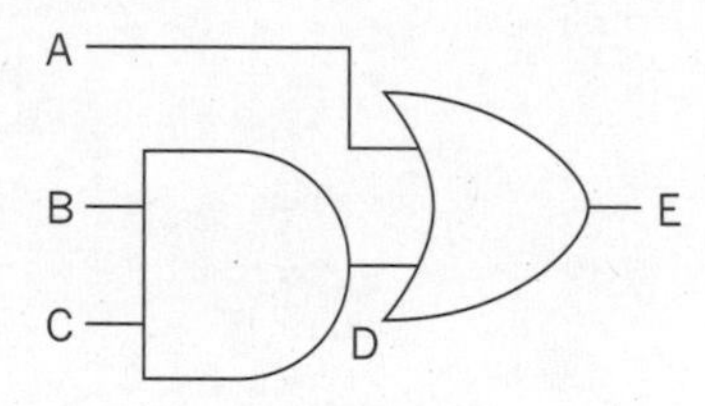

Solving problems with logic or logic circuits means applying skills you have learnt. You may need to combine several different problem-solving activities.

PRACTICE QUESTION

1 You turned these real-life examples into logical expressions in Topic 10.1. Draw a logic circuit to match each of them.

  a Your user name must begin with a capital letter and contain at least one digit.

  b An alarm is sounded if a door is opened and a code is not entered.

  c There are two ways to pass a course: get over 80% in an exam or get 60% and then complete a practical test.

  d To pick up a parcel you must wait two days, then show ID or sign a form.

Invalid expressions

There are rules about how logical operators are used.

- The operator NOT must go in front of a value, making a new value.
- The operators AND and OR must join two values.

If the operators are used in other ways then the Boolean expression is invalid – it doesn't make sense. You can't work out a logical value for the expression if it is not valid.

These Boolean expressions are valid:

A AND NOT B

C AND (A OR B)

NOT (C OR B)

These Boolean expressions are not valid:

A NOT B

OR B

A AND OR C

C AND B NOT

Here are the reasons.

Expression	Why is it not valid?
A NOT B	NOT does not join two expressions
OR B	OR must have an expression on either side
A AND OR C	AND and OR can't be placed next to each other
C AND B NOT	NOT must come in front of an expression not after it

PRACTICE QUESTION

2 Identify which of these expressions are invalid. For each invalid expression explain your reasons.

a AND A NOT B

b NOT A AND NOT C

c NOT C NOT B

d C OR A AND B

STRETCH YOURSELF!

If A and B are True and C is False, work out the truth value of any valid expressions.

SUMMARY QUESTIONS FOR CHAPTERS 8–10

1 Draw diagrams of the three logic gates. Label the inputs and outputs using letters.

a NOT gate

b AND gate

c OR gate

2 To get a drink from a coffee machine the user must put a pound coin in a slot then press a button to select a drink. State which logic gate should be used in the coffee machine to activate the coffee-making feature.

3 Draw a truth table for each of the three logic gates.

a NOT gate

b AND gate

c OR gate

4 Draw a truth table to match this logic circuit.

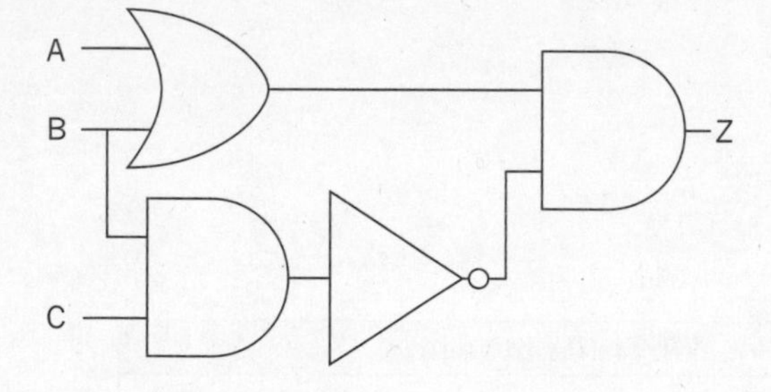

5 Give the truth values of these expressions.

a `12 ≠ 19`

b `"Hello" == "Goodbye"`

c `4 > 8`

d `5 < 5.1`

6 Suppose you know that A is True, B is False, and C is False. Give the truth value of the following expression.

A AND B OR NOT C

7 Only one of these expressions is a valid logical expression. Identify which one.

a B NOT C

b OR B AND C

c A OR NOT B

d NOT A B OR C

8 Draw a truth table for each of the boolean expressions below (assume you do not know the values of A, B, and C).

a (A AND B) OR (A AND NOT C)

b NOT (A OR B) AND C

c NOT A AND NOT B OR C

d A OR NOT B AND C

9 Suppose you do not know the values of A, B and C. Create a truth table for the following Boolean expression.

(A AND B) OR (A AND NOT C)

10 Draw a truth table for each of the expressions.

a $\neg A \wedge B \wedge \neg C$

b $\neg(A \wedge B) \wedge \neg C$

c $\neg A \wedge (B \wedge \neg C)$

d $\neg(A \wedge B \wedge \neg C)$

11 Write the Boolean algebra symbols that correspond to the following operations.

a negation

b conjunction

c disjunction

12 Draw a truth table for the following Boolean expression.

$\neg A \wedge B \vee \neg C$

13 A warning chime goes off in a car if the speed exceeds 5 mph and the seatbelt is not connected. It will also sound if the speed exceeds 5 mph and the door is open. The car has three sensors.

- Sensor A detects if the car is going faster than 5 mph.
- Sensor B detects if the seatbelt is connected.
- Sensor C detects if the door is closed.

a Write a Boolean expression using the operators AND, OR, and NOT and the letters A, B, and C to represent the conditions that control the warning chime.

b Draw a truth table for the expression.

c Draw a logic circuit diagram that could be used to control the warning chime.

d Write an expression in Boolean algebra that corresponds to the warning chime system.

11 MATHS IN ALGORITHMS

11.1 Numerical expressions

Programming and algorithms

An algorithm sets out a series of steps that solve a problem. An algorithm has a similar structure to a program. Programmers often write algorithms when they are planning a program.

In this chapter algorithms will be shown using pseudocode. Pseudocode is a method for writing out algorithms. Pseudocode is similar to a programming language, but it is intended to be read by a human and cannot be understood by the computer.

Variables

The value of an expression can be stored as a named variable. This is called 'assigning a value' to the variable. The variable represents the value that has been assigned to it.

A value can be assigned to a variable using an arrow symbol or an equals sign. For example:

```
game = "Basketball"          or          game ← "Basketball"
```

This assigns a text value to the variable. You can also assign a number value to a variable:

```
age = 18
```

A value can also be assigned using an arrow symbol. This command assigns the result of a calculation to a variable:

```
total ← 17 + 2 + 9
```

You may see either the arrow symbol or the equals symbol used to assign a value to a variable. Both are used in this chapter.

Arithmetic operators

A numeric expression can be a calculation. The result of the calculation is a number value. For example, after this command:

```
total = 17 + 2 + 9
```

the variable 'total' holds the value 28.

Calculations are expressed using arithmetic operators. Here are the four most commonly used operators:

Arithmetic operator	What it does
+	addition
-	subtraction
*	multiplication
/	division

PRACTICE QUESTION

1 Calculate the value assigned to the variable after each of these commands.

a `total = 49 + 19`

b `total = 8 * -9`

c `total = 150 - 177.25`

d `total = 18/4`

Exponentiation

Exponentiation means raising a number 'to the power' of another number.
For example:

two to the power four

In ordinary maths we often write the second number as superscript, like this:

2^4

The meaning of this expression is to multiply the number 2 by itself four times.
In other words:

$2 \times 2 \times 2 \times 2$

In an algorithm exponentiation is written as follows:

```
2^4
```

Precedence

More than one arithmetic operator can be used in an expression.
Calculations are carried out in this order of precedence:

- expressions in brackets
- exponentiation
- multiplication and division (including `DIV` and `MOD`)
- addition and subtraction.

NOTE: A value that is set once at the start of the program and then not altered is called a constant. The constant is given a value in just the same way as a variable, and used in the same way. It represents a value that will not change, such as the value of Pi.

Integers and real numbers

An integer is a whole number, with no decimal place or fractional component. An integer can be a positive or a negative number.

Other number values are called real numbers. If a number has a decimal place in it, it is a real number not an integer. In many programming languages, such as Python, real numbers are called floating point numbers (or float).

The values 19.9 and 10.0 are both real numbers. Both numbers have decimal points, even though 10.0 has a value of zero after the decimal point.

PRACTICE QUESTION

2 Calculate the value assigned to the variable after each of these commands.

a `result ← 7^2` **b** `result ← 5^5`

c `result ← 100^3` **d** `result ← 2^8`

3 Calculate the value assigned to the variable after each of these commands.

a `total = (50 + 20) / 7` **b** `total = 50 + (21/7)`

c `total = 150 - (77/7)` **d** `total = 18/4 + 2.5`

e `total = 8^2 - 4`

4 For each number value state whether it is an integer or a real number.

a 1 000 000 **b** 99.0

c –70 **d** 0.001

REMEMBER: Arithmetic operators are used to transform number values to create numerical expressions.
The exponentiation operator raises one number 'to the power of' another.
7^3 means 'seven to the power three' or $7 \times 7 \times 7$.

11.2 Integer division

Integers and calculations

The value of a calculation might be an integer or a real number. That depends on the operator used in the expression. It also depends on the values used in the expression.

In the case of exponentiation, multiplication, addition, and subtraction:

- if all values are integers, the result will be an integer
- if one or more of the values is a real number, the result will be a real number.

Division and real numbers

The result of division is always a real number rather than an integer. For example, after this command the variable x will have the value 2.5:

```
x = 50/20
```

After this command the variable x will have the value 3.0:

```
x = 60/20
```

PRACTICE QUESTION

1 Go back through all the calculations you did in Topic 11.1. For each answer state whether it is a real number or an integer.

Integer division

Integer division gives an integer result without the fractional component. Integer Division is shown using the operator `DIV`. For example

```
x = 50 DIV 20
```

The result of 50 divided by 20 is 2.5. Take off the decimal point and any numbers that follow it. The variable x has the value 2. An expression using `DIV` always has a value that is an integer.

PRACTICE QUESTION

2 Calculate the value assigned to the variable after each of these commands.

a `x = 100 DIV 3`

b `x = 17 DIV 4`

c `x = 129 DIV 10`

d `x = 44 DIV 5`

Modulus

Most division expressions do not produce a result that is an exact integer. We can show the answer using a decimal point:

50/20 = 2.5

Or using a remainder

50/20 = 2 remainder 10

Modulus means the remainder after an integer division. In an algorithm it is shown using the operator `MOD`. For example:

```
y = 50 MOD 20
```

This expression gives y the value 10. The result of an expression using `MOD` is always an integer.

PRACTICE QUESTION

3 Calculate the value assigned to the variable y after each of these commands.

a `y ← 100 MOD 3`

b `y ← 17 MOD 14`

c `y ← 129 MOD 11`

d `y ← 44 MOD 12`

WORKED EXAMPLE

1 Calculate the value of the variable a after the following command.

```
a = 59 DIV 7
```

The expression uses the operator `DIV`. The operation is integer division. From the seven times table you know:

$7 \times 8 = 56$

$7 \times 9 = 63$

You take the value that is below the given value. In this case:

$7 \times 8 = 56$

This tells us that after integer division by seven, a has the value 8.

2 Calculate the value of the variable b after the following command.

```
b = 59 MOD 7
```

The expression uses the operator `MOD`. The operation is modulus (remainder after division). You already know that the result of integer division is 8.

$7 \times 8 = 56$

To find the remainder subtract this value from the number you started with.

$59 - 56 = 3$

This gives the remainder

`59 MOD 7` is equal to 3

So b has the value 3.

PRACTICE QUESTION

4 Calculate the value assigned to the two variables x and y after each of these pairs of commands.

a `x = 100 DIV 8` `y = 100 MOD 8`

b `x = 17 DIV 14` `y = 17 MOD 14`

c `x = 1234 DIV 2` `y = 1234 MOD 2`

d `x = 99 DIV 5` `y = 99 MOD 5`

> **REMEMBER:** `DIV` is the operator for integer division. A normal division is carried out but only the whole number part of the result is given, leaving out any fractional part or remainder.
>
> `MOD` is the operator for modulus. A normal division is carried out but only the remainder is given.

11.3 Logical expressions

True/False values

You have looked at numerical expressions, which represent number values. Algorithms also include logical expressions, which hold the value True or False.

The values True and False can be assigned to variables, for example:

```
pass = True
```

Logical expressions

Logical values are generated by tests. Logical tests typically compare two values, using relational operators (Topic 9.2). Logical expressions can be combined using Boolean operators to create larger composite logical expressions (Topic 9.3).

WORKED EXAMPLE

Here are two lines from an algorithm:

```
age = 12
adult = (age ≥ 18)
```

For each line state:

a the name of the variable

b the value of the variable

c the data type of the variable.

The first line assigns the value 12 to a variable called age.

a name of variable: age

b value of variable: 12

c data type of variable: integer

The second line creates a new variable called 'adult', which stores the result of a logical test. A logical test compares two values using a relational operator. The relational operator is ≥, which means greater than or equal to. In this case the logical test checks whether the value of age is greater than or equal to 18. As age stores the value 12, this test has the value False.

a name of variable: adult

b value of variable: False

c data type of variable: Boolean

PRACTICE QUESTION

1 An algorithm contains these two commands:

```
passcode = "1234"
cardswipe = False
```

Give the value assigned to the variable x after each of these commands.

a `x ← (passcode = "abcd")`

b `x ← NOT (cardswipe)`

c `x ← (passcode = "1234") OR cardswipe`

d `x ← (passcode ≠ "1234") AND NOT cardswipe`

Selection

Logical expressions are used in selection structures. Selection structures begin with a logical expression, which can have the value True or False. If the expression has the value True then one set of actions has the value carried out. If the expression is False then another set of commands is carried out.

For example:

```
if age < 18 then
    ticketprice = 15.00
else
    ticketprice = 25.00
endif
```

If the test result is True then ticket price is set to 15. If the test result is False then ticket price is set to 25.

PRACTICE QUESTION

2 Here is an algorithm. One line is replaced by ???.

```
x ← input
if (x MOD 2 = 0) then
    print(x, "is an even number")
else
    ???
endif
```

a Give the command you could add instead of the ??? line.

b Write an algorithm that will print a message telling you whether a number can be divided exactly by 10.

c Write an algorithm that inputs two values and prints a message telling you whether the first can be divided exactly by the second.

STRETCH YOURSELF!

Write an algorithm that reports whether a given number is a power of two.

11.4 Iteration

Iteration means repetition. Iteration is also called a loop structure. Many programs include repeated commands. This means that a short algorithm can do a lot of work and process a lot of data.

Every loop must have an exit condition. The exit condition is what stops the loop. There are two types of loop, with different exit conditions.

- A *condition-controlled loop* is controlled by a logical test. When the test has the value False the loop will stop.
- A *counter-controlled loop* is controlled by a counter. When the counter reaches a set value the loop will stop.

Condition-controlled iteration

A condition-controlled loop is controlled using a logical test. There are two common types of condition-controlled loop:

- 'while…' loop
- 'repeat… until' loop.

While… loop

A while loop has the logical test at the top of the loop. If the logical test result is True then the loop will repeat. If the logical test result is False the loop will stop. Here is an example:

```
total = 0
while total < 100 do
    n = input
    total = total + n
endwhile
```

This algorithm repeats the command n = input. This means the user can type in many different values. Each value is added to a total. The logical test is:

```
total < 100
```

If total is smaller than 100 then the loop will continue. If the total is 100 or greater then the loop will stop.

PRACTICE QUESTION

1 An algorithm contains these commands.

```
passcode ← input
while passcode ≠ "1234" do
    print("Error")
    cardswipe ← input
endwhile
```

a The first line takes user input. Give the user input that will prevent the loop from starting.

b Once this loop has begun there is no way to stop this loop from repeating. Explain why not.

c Rewrite the algorithm to include a command inside the loop that will let the user stop the loop by entering the right passcode.

d Rewrite the while command so the loop will also stop if the value 1 is input to the variable cardswipe.

Repeat . . . until loop

A repeat . . . until loop has the logical test at the end of the loop. If the result of the logical test is False then the loop will repeat. If the result of the logical test is True the loop will stop.

Here is an example:

```
total = 0
repeat
    n = input
    total = total + n
until total >= 100
```

This produces the same result as the while loop example shown above.

PRACTICE QUESTION

2 An algorithm contains these commands.

```
attempts ← 0
repeat
    print("What is 7 * 3")
    answer ← input
until ???
```

a The loop is supposed to stop if the user enters the right answer to the question. Complete the last line of the loop.

b The variable 'attempts' must record how many attempts the user made to answer the question. Develop the algorithm so that it does this and outputs the number of attempts at the end.

3 a Convert the algorithm from question 1 to use a repeat . . . until loop.

b Convert the algorithm from question 2 to use a while loop.

For loop

A for loop is a counter-controlled loop. It will typically count up from 0 to a maximum value.

In some programming languages the for loop will stop just before it reaches the maximum value. In other programming languages it will stop when it reaches the maximum value. For example:

```
for i = 0 to 5
    print(i)
next i
```

This would print out the numbers 0 to 5.

The following example would print out the numbers 0, 2, 3, and 4.

```
for i = 0 to 5
    print(2 * i)
next i
```

PRACTICE QUESTION

4 Write a program using a for loop that will print out the number 2 raised to the power 0, 1, 2, 3, and 4.

11.5 Mathematical functions

Subroutines

Many programs include subroutines. A subroutine is a collection of commands. Subroutines are declared and called.

- *Declared*: a series of commands is stored as a subroutine. The subroutine is given a name.
- *Called*: the name of a subroutine is included in a program. When the program runs, the commands stored in the subroutine are carried out.

Using subroutines makes programs shorter and reduces unnecessary repetition of commands.

Procedures and functions

There are two types of subroutine:

- procedures
- functions.

A function creates a new value. The new value is returned to the main program when the function is called. Most programming languages include functions.

Parameters and returned values

A function transforms a value into a new value. The value that goes INTO the function is called a parameter. The value that comes OUT of the function is called a returned value.

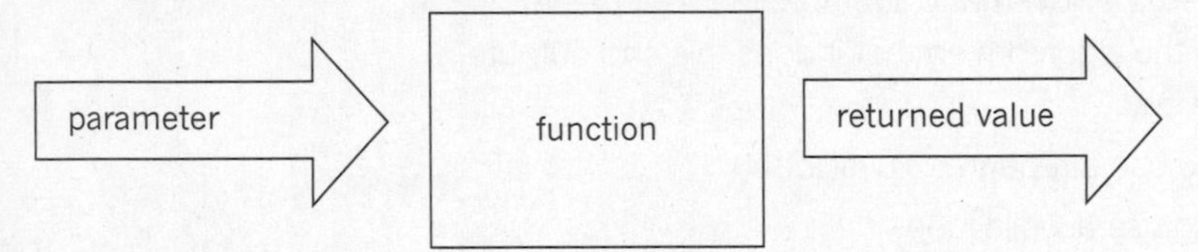

For example, a programmer wrote a function called `is_even` . If an even number was passed into the function it returned the value True. If an odd number was passed into the function it returned the value False.

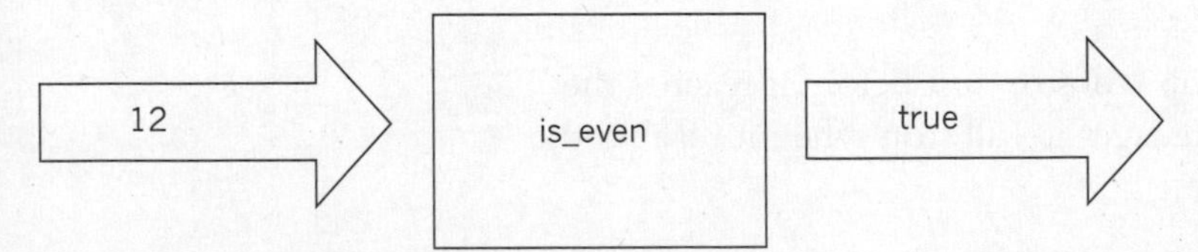

Calling a function

To call a function you enter the name of the function in a program. The parameter of the function is included in brackets after the function name. In this example the function `is_even` is called with the parameter 12.

```
is_even(12)
```

The value returned by the function must be assigned to a variable so that the returned value can be used in the program. In this example the value returned by the function is assigned to a variable called `x`.

```
x = is_even(12)
```

In this example `x` would have the value True.

PRACTICE QUESTION

1 Below are some lines from programs. Each line calls a function. Identify the name of the function, the value of the parameter passed to the function, and the variable that captures the returned value.

a `result = double(19)`　**b** `answer = half(20)`

c `x = factorial(5)`　**d** `final = fraction(6)`

Declare a function

A function stores a series of commands. Declaring a function means writing out the commands. Start by giving the name of the function. In this example the function is called `is_even`.

```
function is_even()
```

You have seen that functions need parameters. The first line of the function must include the name of the parameter. In this example the parameter is stored using the name `number`.

```
function is_even(number)
```

Next the function includes commands to return the required result to the main program. In this case the test result is True if the number is even and False if the number is odd.

```
function is_even(number)
  if (number MOD 2 == 0) then
    return True
  else
    return False
  endif
end function
```

PRACTICE QUESTION

2 Here is a function declaration.

```
function y(z)
  x = 1/z
  return x
end function
```

a State the name of the function.

b Give the name of the parameter.

c If the parameter has the value 4, give the value returned by the function.

d Write a line from a program that calls this function with the value 12 and assigns the returned value to a variable called a.

11.6 Efficiency of algorithms

Processing data structures

A data structure such as a list stores a series of data values. Data structures are typically processed using counter-controlled loops. The loop will count up through the list, processing each item in turn. The number of operations is directly and proportionally related to the number of items in the list. Read more about counter-controlled loops in Topic 11.4.

An example of an algorithm that processes data structures is a linear search. The algorithm processes each data item in turn, comparing it to the search term. The longer the list, the more comparisons are required.

Efficiency of an algorithm

The exact time that the computer takes to process a list will depend on many other factors.

- Is it a fast computer?
- Is the computer doing other work at the same time?
- How many items are in the list?

But even taking all these factors into account, some programs run more quickly than others. They use fewer operations so they run more quickly. Efficient algorithms are algorithms that get a job done with as few operations as possible.

The number of items in a list is called 'n'. The efficiency of an algorithm is shown as a value of n. If the number of operations is directly related to the value of n – that is, if doubling the number of items also doubles the number of operations – this is called linear complexity.

WORKED EXAMPLE

A computer takes 20 seconds to search a list of 5 million items using a linear search. Calculate how long it would take to process a list of half a million items.

Divide the time by five to find out how long it takes the computer to process a million items.

$$\frac{20}{5} = 4$$

The computer takes 4 seconds to process a million items. Therefore, we can calculate that it will take 2 seconds to process half a million items.

PRACTICE QUESTION

1 A computer takes 0.025 seconds to carry out a linear search in a list of 1000. Calculate how long it will take to search lists of the following lengths.

a 300 000 items

b 100 items

c 5 million items

d 72 million items

Express your answer using suitable units (seconds, minutes, or hours).

Nested loops

Some algorithms need to process a list multiple times. This type of algorithm may include a structure called a nested loop. A nested loop is a loop inside a loop.

An example is the bubble sort algorithm.

- The bubble sort must pass down the list. That takes n operations.
- The bubble sort must pass down the list n times.

Remember that n means the number of items in a list. The number of operations is $n \times n$. We can also write this as n^2 or n to the power two.

In some cases a bubble sort will take fewer operations, for example, if the list is already quite well sorted. But when we compare algorithms, it is best to think about the 'worst case'. That is what happens if the list is *not* well sorted already.

Polynomial complexity

The number of operations needed for a linear search is n. The number of operations needed for a bubble sort is n to the power two. So bubble sort takes much longer than linear search.

- If the number of operations is n to the power of another number, this is called polynomial complexity.
- An algorithm with polynomial complexity is much slower than an algorithm of linear complexity.

Algorithms with loops inside loops typically have polynomial complexity.

WORKED EXAMPLE

Calculate how many operations are required to process a list of 30 items using a bubble sort.

A bubble sort has polynomial complexity. The number of operations is n^2

In this case n is 30 so the number of operations is $30 \times 30 = 900$.

PRACTICE QUESTION

2 An algorithm has polynomial complexity. The number of operations required to process a list of n items is n^2. Calculate how many operations are required to process lists of the following lengths.

a 5 items

b 100 items

c 200 000 items

d 8 million items

SUMMARY QUESTIONS FOR CHAPTER 11

1 Give the values of the following numerical expressions:

a `x ← 7 * (9 + 3)`   **b** `x ← (5 + 10)/3`

c `x ← 100 - 6 * 12`   **d** `x ← 4 ^ 3`

2 Give the values of the following numerical expressions.

a `29 DIV 5`   **b** `120 DIV 90`

c `42 MOD 9`   **d** `88 MOD 11`

3 An algorithm contains these two commands.

```
x = 123
y = 100
```

Give the truth values of the following expressions.

a `x == y`   **b** `NOT (x ≤ y)`

c `(y MOD 2) > 0`   **d** `(x > y) AND (y > 90)`

4 Here is an algorithm. Give the values that would be output by the algorithm.

```
total = 0
x = 1
while total < 25 do
  total = total + 2 ^ x
x = x + 1
output (total)
endwhile
```

5 Write a command that will call a function.

a The name of the function is `make_pos`.

b The parameter is –99.

c The returned value will be assigned to a variable called `x`.

6 Write a function called `make_pos`. The function turns negative numbers into positive numbers. For example, the parameter –99 would be returned as 99. If the parameter is positive it is not changed in value.

7 Write an algorithm that allows the user to input a series of values using a condition-controlled loop. The values are added together to give a total. If the user enters a negative number the loop will stop.

8 Write a program that prints out the eight times table, from 8 × 1 to 8 × 12, using a for loop.

9 Here is a function declaration:

```
function final(val)
    digit = val MOD 10
    return digit
end function
```

a Give the name of the function.

b Give the name of the parameter.

c If the parameter has the value 1234, give the value returned by the function.

d In your own words, describe what this function does.

Index

Notes